T0257861

Weed and Pest Control: Myriad Concerns

Weed and Pest Control: Myriad Concerns

Edited by **Jordan Smith**

New York

Published by Callisto Reference,
106 Park Avenue, Suite 200,
New York, NY 10016, USA
www.callistoreference.com

Weed and Pest Control: Myriad Concerns
Edited by Jordan Smith

© 2015 Callisto Reference

International Standard Book Number: 978-1-63239-623-5 (Hardback)

Printed in the United States of America.

Contents

Preface

Extensive information regarding the topic of weed and pest control has been elucidated in this book. The book elucidates different insect control techniques, like tactics in integrated pest management of opportunistic generalist insect species, insect pest control by polyculture technique, the allelopathy phenomenon, application of numerous integrated pest management programs, carbon stocks to manage weeds, biological control of root pathogens, and soil physical procedures and irrigation tactics.

This book is a result of research of several months to collate the most relevant data in the field.

When I was approached with the idea of this book and the proposal to edit it, I was overwhelmed. It gave me an opportunity to reach out to all those who share a common interest with me in this field. I had 3 main parameters for editing this text:

1. Accuracy – The data and information provided in this book should be up-to-date and valuable to the readers.
2. Structure – The data must be presented in a structured format for easy understanding and better grasping of the readers.
3. Universal Approach – This book not only targets students but also experts and innovators in the field, thus my aim was to present topics which are of use to all.

Thus, it took me a couple of months to finish the editing of this book.

I would like to make a special mention of my publisher who considered me worthy of this opportunity and also supported me throughout the editing process. I would also like to thank the editing team at the back-end who extended their help whenever required.

Editor

An Overview of Chilli Thrips, *Scirtothrips dorsalis* (Thysanoptera: Thripidae) Biology, Distribution and Management

Vivek Kumar, Garima Kakkar, Cindy L. McKenzie,
Dakshina R. Seal and Lance S. Osborne

Additional information is available at the end of the chapter

1. Introduction

The industrial revolution, globalization and international trade liberalization are some of the important events that have afforded vast opportunities for invasive insect species to establish in new territories [1]. These invasive species, facing no challenge by their natural enemies, thrive well in the new environment [2]. In addition to the disturbance they cause to the biodiversity, pest invasion in any country results in increased pressure on biosecurity, national economy, and human health management systems [1, 3, 4]. Apart from economic loss in managing them, these pests pose a significant detrimental impact on tourism and recreational value of the region, which further adds in indirect economic damage to the nation [5]. Of this large group of invasive pests, thrips are one of the most important members. The invasive status gained by thrips across the globe is due to their high degree of polyphagy, wide host range and easy dispersal that can be anthropogenic or natural (wind-mediated).

The earliest fossil record of order Thysanoptera dates back to the Late Triassic period, from the state of Virginia in the United States and the country Kazakhstan in Central Asia, but their abundance was rare until the Cretaceous period from which many specimens of Thysanoptera have been recorded [6]. The order Thysanoptera was given its current taxonomic rank by an Irish entomologist, A. H. Haliday in 1836, and since then more than 8,000 species of thrips have been reported. In this insect order, the genus *Scirtothrips* Shull contains more than 100 thrips species, among which 10 species have been reported as serious pests of agricultural crops [7]. Within this genus, *Scirtothrips dorsalis* Hood is a significant pest of various economically important vegetable, ornamental and fruit crops in southern and eastern Asia, Oceania and parts of Africa [8, 9]. *S. dorsalis* is native to the Indian subcontinent and is a polyphagous pest with more than 100 reported hosts among 40 different families of plants [10]. However, in the past two decades, increased globalization and open agricultural trade has resulted in the vast expansion of the geographical distribution and host range of the pest. In the United States, it is a new invasive pest where the first established population of *S. dorsalis* was reported in 2005 from Florida. Since then it has emerged as a serious pest of various economically important host crops in the southeastern regions of the United States. It has been reported from 30 counties in Florida, 8 counties in Texas with several positive reports of its invasion from Alabama, Louisiana, Georgia, and New York. In a recent study [11], this pest was found attacking 11 different hosts at a fruit nursery in Homestead, Florida. Interestingly, they were found to reproduce on nine plant taxa that had not previously been reported as host plants in the literature suggesting that the host range of this insidious pest is continuing to expand as it invades new regions. The small size and cryptic nature of adults and larvae enables *S. dorsalis* to inhabit microhabitats of a plant often making monitoring and the identification difficult. *S. dorsalis'* life stages may occur on meristems and other tender tissues of all above ground parts of host plants [12]. Consequently, the opportunity of trans-boundary transportation of *S. dorsalis* through the trade of plant materials is high [13]. Existence of any variation in phenotypic and genetic makeup of such a pest makes identification much more difficult [14].

This chapter is intended to summarize the parameters facilitating worldwide distribution of this pest, damage potential and the advancement in the post-invasion management strategies being practiced in the United States and other parts of the world. The focus will be on the latest development in the integrated pest management of *S. dorsalis* including identification techniques and biological, chemical and cultural control strategies.

2. Background information

The great reproductive potential and keen ability for invasion combined with easy adaptation to newly invaded areas are a few of the qualities which make *Scirtothrips* species major concerns for agriculture in many countries [15]. From the beginning, *S. dorsalis* has been reported as an opportunistic generalist species that is able to feed on a variety of host plants, depending upon availability in the region of incidence. The first reference to *S. dorsalis* was in early 1900's when it was reported damaging the tea crop in the Tocklai area of Assam state in India. In later years *S. dorsalis* was responsible for damaging the tea crops in all of the major tea growing regions of eastern India including Cachar, the Assam Valley, Terai and the Dooars [16]. In 1916, this pest was reported infesting castor in the Coimbatore district of the southern part of India and later was found infesting other hosts in the region including chilli, groundnuts, mango, beans, cotton, brinjal (eggplant) and *Casia fistula* [17, 18]. Young leaves, buds, and tender stems of the host plants were severely damaged. Thrips repeated puncturing of tender leaf tissues with their stylet produces 'sandy paper lines' on the epidermis of the leaves and eventual crinkling of leaves. In India, the characteristic leaf curl damage caused by this pest is known as "Murda" (Hindi meaning- dead body) disease, because infestations resulted in the death of plants [19]. Many different scientific names have been assigned to *S. dorsalis* since it was first described in

1919, mainly because of the lack of sufficient scientific literature regarding morphological differences and variations in host range from the different geographical regions. During the last 100 years, the host range and the bio-geographical range of *S. dorsalis* have broadened. The thrips is established in all of the habitable continents except Europe, where repeated introductions have been intercepted and eliminated [13]. Studying the history of *S. dorsalis* aids in the understanding of behavioral and morphological diversity exhibited by this species as a result of biological and ecological variations that have occurred during its long migration to different parts of the world.

3. Geographical distribution

3.1. Worldwide distribution

S. dorsalis is widely distributed along its native range in Asia including Bangladesh, Brunei Darussalam, China, Hong Kong, India, Indonesia, Japan, Republic of Korea, Malaysia, Myanmar, Pakistan, Philippines, Sri Lanka, Taiwan, and Thailand. Further south *S. dorsalis* occurs in northern Australia and the Soloman Islands. On the African continent, the pest is reported from South Africa and the Ivory Coast, with plant health quarantine interceptions suggesting a wider distribution across West Africa and East Africa (Kenya) [20]. *S. dorsalis* is in Israel as well as in the Caribbean including Jamaica, St. Vincent, St. Lucia, Barbados and Trinidad [12]. In South America, *S. dorsalis* has been found causing serious damage to grapevine in western Venezuela [20].

3.2. U.S. invasion

Changing climatic conditions and globalization have resulted in the increasing importance of invasive species as recurrent problems around the globe. More than 50,000 non-indigenous species have already been introduced into the United States, causing an estimated annual damage of more than $120 billion in forestry, agriculture and other sectors of society [3, 21]. The rich vegetation and neotropical climate of Florida make the state suitable for the invasion and establishment of exotic flora and fauna [22]. *S. dorsalis* is a newly introduced insect pest in Florida believed to have originated from Southeast Asia. In between 1984-2002, it was intercepted about 89 times by USDA-APHIS inspectors at various US ports-of-entry [23]. Most of the records of interception were from imported plant materials including cut flowers, fruits and vegetables. With the exception of Hawaii, the presence of this tropical south Asian pest was not confirmed in the Western Hemisphere until 2003. In Florida, *S. dorsalis* was reported from Okeechobee County in 1991 and from Highland County in 1994 but failed to establish a durable population [24]. In 2003, Tom Skarlinsky (USDA-APHIS-PPQ) reported live larvae and pupae under the calyx of treated peppers in a shipment of *Capsicum spp.* traced back to hot pepper production areas in St. Vincent and the Grenadines, West Indies [25]. Later, with the collaborative efforts of the USDA (APHIS) and the Institute of Food and Agricultural Sciences (University of Florida), *S. dorsalis* was found established in different agricultural districts of St. Lucia and St. Vincent [26], Barbados, Suriname, Trinidad and Tobago, and

Venezuela [25]. In 2005, *S. dorsalis* was found on pepper and 'Knockout' Rose plants in retail garden centers in Florida and Texas. Subsequently, *S. dorsalis* has been reported many times on different ornamental plants in commercial nurseries throughout Florida [27]. In a collaborative survey over a two-month period (Oct-Nov 2005), the Florida Department of Agricultural and Consumer Services (FDACS) and the University of Florida found infestations 77 times in 16 counties [25]. Of the 77 positive observations, 66 were found on roses, 10 on *Capsicum* and one on *Illicium*.

Venette and Davis [28] projected the potential geographic distribution of *S. dorsalis* in North America. Based on this *S. dorsalis* could extend from southern Florida to the Canadian border, as well as to Puerto Rico and the entire Caribbean region which suggests that this pest could also become widely established in South America and Central America. The small size (< 2 mm in length) and thigmotactic behavior of *S. dorsalis* make it difficult to detect the pest in fresh vegetation, thus, increasing the likelihood of the transportation of the pest through international trade of botanicals. The major pathways of trans-boundary movement of *S. dorsalis* includes (i) air passengers and crew, their baggage, and air cargo of plant propagative materials and fresh ornamentals, fruits, and vegetables, (ii) mail, including mail from express mail carriers, (iii) infested smuggled fresh plant materials, and (4) windborne dispersal [29].

4. Economic impact

Among 8,800 species of thrips, around 5,000 species has been well described with their diverse life history and habitats [6]. Approximately 1% of the members of this order have been reported as serious pests by humans owing to various damages which disrupt their life styles [30]. Thrips can reduce yield or value of the crop directly by using them as food and oviposition site and indirectly by transmitting plant diseases. Their infestation can negatively impact global trade due to the quarantine risks associated with several species in the order. The majority of scientific literature related to economics of thrips deals with four important thrips species: *Thrips tabaci* Lindeman, *T. palmi* Karny, *Frankliniella occidentalis* (Pergande) and *S. dorsalis* [30].

India is one of the world's largest chilli (*Capsicum annum* L.) producers which contributes about 36% (0.45 million tons) of global production [31]. According to a survey by the Asian Vegetable Research and Development Committee, *S. dorsalis* is one of the most important limiting factors for the chilli production in the country along with aphid species *Myzus persicae* Sulzer, *Aphis gossypii* Glover and mite *Polyphagotarsonemus latus* Banks [32]. Yield loss solely dedicated to *S. dorsalis* damage can range between 61 to 74% [33]. Because of its damage potential to chilli pepper, this dreadful pest is commonly referred as chilli thrips.

Globally, the popularity and demand for mango (*Mangifera indica* L.) and its processed product is rising which has resulted in the expansion of area under mango cultivation. Asian countries contribute around 77% of the global mango production followed by Americas (13%) and Africa (9%). In 2005, world mango production was reported as 28.5 million metric tons [34]. Malaysia, which is a major mango consumer (10th largest mango importer in the world), produces mango on over 4,565 ha [35]. However, their domestic mango production has been reported to suffer considerable economic losses due to thrips infestation with the major thrips species responsible for damage of mango panicle reported as *Thrips hawaensis* Morgan and *S. dorsalis* [36]. Along the same international theme, *S. dorsalis* is considered as a major economic threat to grape and citrus production in Japan [37, 38] and vegetable production in China and US [39].

The invasion of *S. dorsalis* into the Caribbean region prompted an economic analysis to be conducted in 2003 by the United States Department of Agriculture on 28 potential hosts of the pest which suggested that a 5% loss of these crops may lead to a loss value of $3 billion to the US economy [40]. Assessment of the damage potential of *S. dorsalis* from Florida's perspective as the port of entry showed that there is an immediate need for development of effective management practices against this pest. In 2010, the US horticulture (greenhouse/nursery) industry contributed approximately $15.5 billion to the US economy, among which Florida was the second largest contributor after California by adding 11.2% to the economy [41]. Florida received cash receipts of approximately $7.80 billion in 2010 from all agricultural commodities among which top three contributors were greenhouse/nursery, orange and tomato which added about 53% of all cash receipts. Strawberries ($362 million), peppers ($295 million), peanuts ($89 million), cucumbers ($88 million), cotton ($49 million), and blueberries ($47 million) contributed an additional 1 billion (approx.) to Florida's economy. Since these crops have potential to serve as host plants of *S. dorsalis*, even 10% loss of these commodities can cause significant impact on Florida's economy and may open the market for foreign trade [42]. Florida Nurserymen and Growers Association consider *S. dorsalis* as one of the thirteen most dangerous, exotic pests threatening the industry [43].

5. Host plants

Prior to the introduction of *S. dorsalis* into the New World, the host range of this pest included more than 100 plant taxa among 40 families [10]. Subsequent to the introduction of *S. dorsalis* into the New World, the pest was found to attack additional taxa of plants [28]. The main wild host plants belong to the family Fabaceae, which includes Acacia, Brownea, Mimosa and Saraca. In its native range of the Indian subcontinent, chilli crops are reported to be attacked by 25 different pests, among which *S. dorsalis* is considered as one of the most serious threats [44]. *S. dorsalis* is also abundant on *Arachis* in India [45], sacred lotus in Thailand [10], and tea and citrus in Japan [46]. Among the potential economic hosts of this pest listed by Venette and Davis [28] are banana, bean, cashew, castor, citrus, cocoa, corn, cotton, eggplant, grapes, kiwi, litchi, longan, mango, melon, peanut, pepper, poplar, rose, strawberry, sweet potato, tea, tobacco, tomato, and wild yams (*Dioscorea spp.*). Interestingly, *S. dorsalis* is not reported reproducing on all of the hosts mentioned in the literature and plant species has been designated as a host plant based on the presence of adult thrips. While *S. dorsalis* is known to forage on wide range of plant species, a true host must be identified by its ability to support thrips reproduction in addition to provisioning food and shelter. Based on information obtained from the Global Pest and Disease Database [47], *S. dorsalis* was reported to feed on (not necessarily reproduce on) more than 225 plant taxa worldwide in 72 different families and 32 orders of plants. In Florida, S. *dorsalis* has been reported from 61 different plants till 2011 (Table 1). Disparities in host selection in different geographical regions are documented in the literature. For example, S. *dorsalis* is reported on mango in Puerto Rico but not in adjacent Caribbean islands where it was reported earlier on other host plants. S. *dorsalis* is a significant pest of citrus in Japan [48] and Taiwan [49], but not in India or the United States. Many factors could be attributed to the differences in host plants of S. *dorsalis* reported from different geographical regions. These various factors could include variation in competition with other pests, availability of predators in the region of invasion, availability of hosts, environmental conditions, etc. [42], but could also be the result of differential biological activity of different S. *dorsalis* biotypes/cryptic species, none of which have yet been reported.

Scientific name	Common or trade name
Antirrhinum majus L.	Liberty Classic White Snapdragon
Arachis hypogaea L	Peanut or groundnut
Begonia sp.	Begonia
Breynia nivosa (W. Bull) Small	Snow bush, snow-on-the-mountain
Camellia sinensis (L.) Kuntze	Tea
Capsicum annuum L.	Jalapeno pepper, Bonnet pepper
Capsicum frutescens L.	Chilli pepper
Capsicum spp.	
Celosia argentea L.	Celosia – red fox
Citrus spp.	
Concocarpus erectus	
Coreopsis sp.	Tickseed
Cuphea sp.	Waxweed, tarweed
Duranta erecta L.	golden dewdrop, pigeonberry, skyflower
Euphorbia pulcherrima Willd.	Poinsettia
Eustoma grandiflorum (Raf.) Shinn.	Florida Blue Lisianthus
Ficus elástica 'Burgundy' Roxb. Ex Hornem.	Burgundy Rubber Tree
Gardenia jasminoides J. Ellis	Jasmine
Gaura lindheimeri Engelm. & Gray	Lindheimer's beeblossom
Gerbera jamesonii H. Bolus ex Hook. F.	Gerber daisy
Glandularia x hybrida (Grönland & Rümpler) Neson & Pruski	Verbena
Gossypium hirsutum L.	Cotton
Hedera helix L.	English Ivy
Illicium floridanum Ellis	Florida anisetree

Scientific name	Common or trade name
Impatiens walleriana Hook. F.	Super Elfin White
Jasminum sambac (L.) Ait.	Pikake
Lagerstroemia indica L.	Crape myrtle
Laguncularia recemosa (L.) Gaertn. f.	White buttonwood
Ligustrum japonicum Thunb.	Japanese privet
Litchi chinensis Sonn.	Litchi
Mahonia bealei (Fortune) Carrière	Leatherleaf mahonia
Manilkara zapota(L.) D. Royen	Sapodilla
Mangifera indica L.	Mango
Murraya paniculata (L.) Jack	Orange-jasmine
Ocimum basilicum L.	Sweet Basil
Pelargonium x hortorum Bailey	Geranium
Pentas lanceolata (Forssk.) Deflers	Graffiti White
Persea americana Mill.	Avocado
Petunia x hybrida	Petunia Easy Wave Red
Pittosporum tobira (Thunb.) Ait. f.	Variegated Pittosporum
Plectranthus scutellarioides (L.) R. Br.	Coleus
Pouteria campechiana (Kunth) Baehni	Canistel
Rhaphiolepsis indica(L.) Lindl. ex Ker Gawl.	Shi Ban Mu
Ricinus communis L.	Castor Bean
Rhaphiolepis umbellate (Thunb.)	Yeddo Hawthorn
Richardia brasiliensis Gomes	Brazil Pusley
Rhododendron spp.	Azalea
Rosa X 'Radrazz'	'Knockout®' rose
Salvia farinacea Benth.	Victoria blue
Schefflera arbicola (Hayata) Merr.	Dwarf umbrella tree
Strobilanthes dyerianus Mast.	Persian shield
Synsepalum dulcificum (Schumach. & Thonn.) Daniell	Miracle fruit
Tagetes patula L.	Marigold
Tradescatia zebrina hort. ex Bosse	Wandering jew
Vaccinium corymbosum L.	Highbush blueberry
Viburnum odoratissimum var. *awabuki* (K. Koch) Zabel	Sweet viburnum
Viburnum suspensum Lindl.	Viburnum
Viola x wittrockiana Gams	Wittrock's violet
Vitis vinifera L.	Grapevine
Zinnia elegans Jacq.	Zinnia Profusion White

Table 1. Confirmed plant hosts of *Scirtothrips dorsalis* in Florida. Source: [80].

6. Host damage

S. dorsalis feeding on the meristems, terminals and other tender plant parts of the host plant above the soil surface results in undesirable feeding scars, distortion of leaves, and discoloration of buds, flowers and young fruits. The pest prefers young plant tissue and is not reported to feed on mature host tissues. The piercing and sucking mouthparts of *S. dorsalis* can damage the host plant by extracting the contents of individual epidermal cells, leading to the necrosis of tissue. The color of damaged tissue changes from silvery to brown or black. The appearance of discolored or disfigured plant parts suggests the presence of *S. dorsalis*. Adults and larvae of *S. dorsalis* suck the cell sap of the leaves, causing the leaves to curl upward [50]. Severe infestations of *S. dorsalis* cause the tender leaves and buds to become brittle, resulting in complete defoliation and yield loss. For example, heavy infestations of pepper plants by *S. dorsalis* cause changes in the appearance of plants termed "chilli leaf curl" [51]. On many hosts, the thrips may feed on the upper surfaces of leaves when infestations are high. Infested fruits develop corky tissues [52]. Sometimes, plants infested by *S. dorsalis* appear similar to plants damaged by the feeding of broad mites. Plants infested with *S. dorsalis* may show the following damage symptoms: (i) silvering of the leaf surface, (ii) linear thickening of the leaf lamina, (iii) brown frass markings on the leaves and fruits, (iv) grey to black markings on fruits, often forming a distinct ring of scarred tissue around the apex and (v) fruit distortion and premature senescence and abscission of leaves [53]. Apart from causing direct damage to its host *S. dorsalis* also vectors seven plant viruses including chilli leaf curl virus (CLC), peanut necrosis virus (PBNV), peanut yellow spot virus (PYSV), tobacco streak virus (TSV), watermelon silver mottle virus (WsMoV), capsicum chlorosis virus (CaCV) and melon yellow spot virus (MYSV) [8, 10, 54, 55, 56, 57].

7. Identification

Correct identification is a primary step in the development of sound management practices against a pest. Identification helps in attaining previously reported information against the subject species [58] crucial in planning and implementation of an appropriate biological research strategy. Morphological identification characters of *S. dorsalis* are well studied in the literature due to its economic importance and global distribution. However, due to the small size and morphological similarities within the genus, the identification of unknown specimen to species level is a challenge to non-experts.

Larvae of *S. dorsalis* are creamish white to pale in color. Sizes of the first instars, second instars, and pupae range between 0.37-0.39, 0.68-0.71 and 0.78-0.80 mm, respectively [12]. Morphological identification of *S. dorsalis* larva can be made using the following features [59]): D1 and D2 setae present on the head and abdominal terga IX of larvae are simple and funnel-shaped, respectively. The D1 setae on terga X are funnel shaped. The larval pronotum is reticulated and has 6-7 pairs of pronotal setae. Abdominal segments IV-VII of larvae have a total of 8-12 setae each. The distal two thirds of the fore-femora of larvae possess four funnel shaped setae and the body of larvae possesses granular plaques.

The body of adult *S. dorsalis* is pale yellow in color and bear dark brown antecostal ridges on tergites and sternites. Adults are less than 1.5 mm in length with dark wings. The head is wider than long, bearing closely spaced lineations and a pair of eight segmented antennae with a forked sensorium on each of the third and fourth segments. Dark spots that form incomplete stripes are seen dorsally on the abdomen [12]. Three pairs of ocellar setae are present, the third pair, also known as the interocellar setae (IOS), arises between the two hind ocelli (HO) and is nearly the same size as the two pairs of post ocellar setae (POS) on the head. The pronotum consists of closely spaced horizontal lineation. The pronotal setae (anteroangular, anteromarginal and discal setae) are short and approximately equal in length. The posteromarginal setae-II is broader and 1.5 times longer than the posteromarginal setae-I and III. The posterior half of the metanotum presents longitudinal striations; medially located metanotal setae arise behind the anterior margin and campaniform sensilla are absent. Three discal setae are located on the lateral microtrichial fields of the abdominal tergites and the posteromarginal comb on VIII segment is complete. The shaded forewings are distally lighter in color with posteromarginal straight cilia on the distal half and the first and second veins bear three and two widely spaced setae, respectively. Discal setae are absent on sternites and sternites are covered with rows of microtrichia, excluding the antero-medial region [60, 61].

Using traditional taxonomic keys, adult thrips can be identified to genus, but due to the intraspecific morphological variations in many species, identifying them to species requires substantial expertise [7]. For many taxa of thrips it is impossible to assign an immature to a particular species in the absence of adults [62]). In addition, high levels of variation in the basic biology, life history, host selection, pest status, vector efficiency and resistance to insecticides exist in different thrips species. Misidentification of thrips species can lead to the misapplication of management practices, resulting in wasted money, resources and time [63]. Selection of the wrong biological control agents due to the ambiguous identification of the target pest discourages growers to adopt chemical free pest management strategies. Thus, a rapid, species-specific, developmental-stage non-limiting method for identification of thrips species is of paramount importance to implement appropriate IPM strategies.

Taxonomic characterization of thrips, including *S. dorsalis*, has always been difficult due to their small size and cryptic nature. Thus, it is important to utilize the advantage of other methods of identification including molecular techniques which is not limited by the factors associated with morphological identification [64, 65]. Molecular techniques can be cost effective, rapid, and performed by non-taxonomic experts. Recently a molecular marker (rDNA ITS2) has been developed for species specific identification of *S. dorsalis* specimens [65]. However, misidentification of specimens using solely molecular identification based on genetic information available in databases such as Genbank and EMBL is very common [66] until a voucher specimen or photo-documentation is available to confirm the identity. Thus, it is important to integrate both identification methods (morphological + molecular) to achieve a double confirmation system for validating identification of various thrips species using a single specimen. There are a few such techniques available such as sonication of specimens for DNA extraction [67] and the automated high-throughput DNA protocol [66], which allows DNA extraction to be performed without destroying the specimen. Another integrated technique available for thrips identification involves piercing the abdominal region of the thrips specimen using a minute pin and processing the extracted gut content for molecular identification prior to the slide mount to preserve the voucher specimen [7]. However, this method requires great skill to keep the specimen intact and save the specimen for slide preparation. Because thrips are soft-bodied minute insects, specimens can be damaged while puncturing the abdomen or during slide preparation. Recently, a new integrated identification technique has been developed for correct identification of thrips using a single specimen. Prior to the DNA extraction of thrips larvae or adults, specimens are subjected to traditional morphological identification using high resolution scanning electron microscopy (SEM) and then gold/palladium sputter coated thrips specimens are processed for polymerase chain reaction assay for molecular identification [14]. Photo-documentation can be easily created for any future reference for the specimen understudy. This novel technique has advantages over other integrated methods as it is simple and quick, utilizes fewer specimens for identification, provides high yield of DNA and can be easily mastered by non-experts.

8. Life cycle

Thysanopterans have always been recorded as opportunistic species, as their life history strategies were preadapted from a detriophagous ancestral group developed in a habitat where optimal conditions of survival were brief [68]. Mating does not result in fertilization of all the eggs and unfertilized eggs produce males while fertilized eggs produce females. Sex ratio is in favor of female progeny [16]. The stages of the life cycle of S. dorsalis include the egg, first and second instar larva, prepupa, pupa and adult. Gravid females lay eggs inside the plant tissue (above the soil surface) and eggs hatch between 5-8 days depending upon environmental conditions [12, 16]. Larvae and adults tend to gather near the mid-vein or borders of the damaged portion of leaf tissues. Pupae are found in the leaf litter, on the axils of the leaves, and in curled leaves or under the calyx of flowers and fruits. Larval stages complete in 8-10 days, and it takes 2.6-3.3 days to complete the pupal stages. The life span of S. dorsalis is considerably influenced by the type of host they are feeding. For example, it takes 11.0 days to become an adult from first instar larva on pepper plants and 13.3 days on squash at 28°C. S. dorsalis adults can survive for 15.8 days on eggplant but 13.6 days on tomato plants [25]. They can grow at minimal temperatures as low as 9.7°C and maximum temperatures as high as 33.0°C. Their thermal requirement from egg to egg is 281-degree days and egg to adult is 265-degree days [25]. Populations are multivoltine in temperate regions with up to eight generations per year and 18 generations per year in warm subtropical and tropical areas [69]). In Japan, S. dorsalis start egg laying in late March or early April when temperatures are favorable for development (70) and first generation adults can be seen from early May [71]. However, S. dorsalis cannot overwinter in regions where temperature remains below -4°C for five or more days [69]. Prolonged rainy seasons do not appear to affect populations much, but the population remains more abundant during prolonged dry conditions than in moist rainy periods.

9. Management strategies

Incursions of *S. dorsalis* are difficult to manage and successful eradication is possible only with early detection and immediate implementation of management practices. Host crops, which develop from seeds such as bean, corn or cotton, must be carefully monitored during the seedling stage of growth because this stage is extremely susceptible to attack by *S. dorsalis* [12]. Symptoms of infestations of *S. dorsalis* must be monitored on their susceptible host plants like roses, pepper, cotton, etc. twice per week and if symptoms appear, then thrips samples should be sent to a reputable laboratory for identification.

9.1. Sampling plan

Appropriate methodology for sampling *S. dorsalis* populations is essential to understand presence and absence of thrips and to determine levels of population abundance at a given time of infestation in a specific host crop. The sampling method has to be economically sound and it should provide information on pest abundance with a minimal number of samples collected. Thus, it is important to determine the within- plant and spatial distribution of the pest in order to select an appropriate sampling unit. For example, melon thrips (*Thrips palmi* Karny) appears on the bottom leaves of most of its vegetable hosts, but on the top leaves of pepper plants. *S. dorsalis* attacks all above ground parts of its hosts, although initiation of infestation can invariably be seen on the young leaves of seedlings and mature plants. As plants grow older, *S. dorsalis* populations may disperse on the entire plant with the abundance on the younger leaves. In a study conducted in St. Vincent [52], *S. dorsalis* developmental stages were observed on all above-ground parts of `Scotch Bonnet' pepper, *Capsicum chinense* Jacq.' in rainy and dry seasons (Table 2). Mean numbers of *S. dorsalis* adults and larvae were most abundant on the top leaves followed by middle leaves and bottom leaves, flowers and fruits. No significant difference was observed in *S. dorsalis* adults and larvae counts reported on the bottom leaves, flowers and fruits.

In general, insects may have clumped, random or regular distribution in the field and at the initial stage of invasion, insects may appear at a certain location(s) of a crop field depending on environmental factors. These locations may be at the edge of the fields or inside the fields. Known factors that influence such distribution includes wind direction, light intensity, soil fertility, soil moisture, crop vigor and crop nitrogen levels. In several of our studies, *S. dorsalis* displayed various patterns of within-field distribution. The distribution patterns of *S. dorsalis* adults in 2004 and 2005 in a pepper planting were either random or regular in the smaller plots (6, 12 and 24 m^2). However, the distribution of adults in the larger plots (48 m^2) was aggregated in October 2004 (rainy season), and regular in March 2005 (dry season). Characterizing hot spots (region of aggregation in a field) helps develop an economical sampling methodology and adoption of site selected management strategies using biocontrol agents, lower volumes of insecticides, and effective cultural control practices.

Direct methods of *S. dorsalis* sampling involves counting thrips on any part of a host plant (e.g. leaf, flower and fruit) by using a hand lens, microscope or the naked eye. In this method, the

Location on	Mean number of Scirtothrips dorsalis		
Pepper plant	Adults	Larvae	Total
Field 1 (October 2004, rainy season)			
Top leaf	4.50a	5.50a	10.00a
Middle leaf	1.75b	2.00b	3.75b
Bottom leaf	0.50b	0.75c	1.25c
Flower	0.75b	0.25c	1.00c
Fruit	0.25b	1.00bc	1.25c
Field 2 (March 2005, dry season)			
Top leaf	2.25a	4.25a	6.50a
Middle leaf	1.00ab	2.25ab	3.25b
Bottom leaf	0.25b	0.75bc	1.00c
Flower	0.50b	0.25c	0.75c
Fruit	0.50b	0.75bc	1.25c
Field 3 (March 2005, dry season)			
Top leaf	3.75a	4.00a	7.75a
Middle leaf	1.25b	1.75ab	3.00b
Bottom leaf	0.75b	0.50bc	1.25bc
Flower	0.25b	0.25c	0.50c
Fruit	0.50b	1.00bc	1.50bc

Means within a column for each field followed by the same letter do not differ significantly ($P > 0.05$, Waller-Duncan k ratio procedure).

Table 2. Within plant distribution of *Scirtothrips dorsalis* adults and larvae on `Scotch Bonnet' pepper plants in three fields in St. Vincent based on samples taken during October 2004 (Field 1), March 2005 (Fields 2 and 3). Source: [52].

part of the plant host sampled may be detached or left intact on the plant. In a beat pan or beat board method, the plant part is tapped against the board to separate *S. dorsalis* adults. More accurately *S. dorsalis* can be sampled by washing plant parts with 70% ethanol or kerosene oil. The contents of the liquid are sieved through a 300 mesh sieve to separate thrips which are then observed using a microscope or hand lens. In an indirect method of *S. dorsalis* sampling, sticky cards of various colors can be placed inside, outside or at the perimeter of the crop field at the level of crop canopy. *S. dorsalis* are attracted to the color and get stuck. Sticky cards can be used from planting to harvest of a crop to monitor thrips advent and abundance during the crop season. Yellow colored sticky cards are commonly used to monitor *S. dorsalis*, but blue, white and green colored cards also attract *S. dorsalis* adults. In a recent study [72] conducted in Taiwan and St. Vincent, three different sticky cards (blue, yellow and white) were evaluated for sampling of *S. dorsalis* and the results suggested that yellow sticky cards could be used efficiently for population detection and monitoring purposes of this pest. In Japan, yellowish-green, green and yellow sticky boards were found to be effective in attracting *S. dorsalis* adults [73]. Irrespective of colors, sticky cards should be replaced every 7-10 days by a new one.

9.2. Cultural practices

Development of effective management practices for *S. dorsalis* is still in its infancy. The World Vegetable Center has several recommendations which could serve as a basic management practice template for the control of this pest in vegetable production. It involves crop rotation, removal of weeds (which may serve as hosts or virus reservoirs), insecticide rotation and supporting the maximum use of natural enemies including predators and parasites. In some of the plant cultivars resistance to *S. dorsalis* feeding appears to exist. Presence of gallic acid plays a crucial role in resistance to *S. dorsalis* in some varieties of the pepper plant [25]. Recently, researchers at the Mid-Florida Research and Education Center, University of Florida, screened for plant resistant to *S. dorsalis* feeding in 158 different cultivars of pepper and found 14 of these cultivars were resistant to the pest attack. "Brigadier hybrid" and "Trinidad perfume" were among the highly resistant cultivars.

In Japan, synthetic reflective (vinyl) film has been used to protect citrus crops from *S. dorsalis* infestations [74]. In another study, the use of white aqueous solution, i.e. 4% $CaCO_3$ on mandarin orange trees along with reflective-sheet mulching reported to provide effective suppression of *S. dorsalis* populations [75]. Common cultural practices like vermiwash in addition to the use of vermi-compost and neem (*Azadirachta indica* A. Juss.) cake has also been found effective in regulating *S. dorsalis* attack on pepper [76].

9.3. Chemical control

Chemical control is the primary mode of management of *S. dorsalis* and a wide range of insecticides belonging to different chemical groups is currently used worldwide to control this pest. In south-central Asia, chemical control is conducted using older chemistries including organophosphates such as quinalphos, dimethoate, and phosphamidon as well as the carbamate, carbaryl. In India and Japan monocrotophos, also an organophosphate and the pyrethroid permethrin gave better suppression of this pest (50, 77). Organophosphates (malathion and fenthion) were also found effective against *S. dorsalis* on grapevine [78]. Since their introduction in the Greater Caribbean, there was a lack of information for effective management of this insect using modern insecticides. In recent years, effectiveness of various novel chemistries against *S. dorsalis* has been evaluated and 10 chemical insecticides belonging to seven different modes of action classes (Table 3) have been reported to provide good control of the pest [79, 80]. The rotational use of three or more insecticides from different action classes have been suggested to get prolonged suppression of the pest population [81]. Pyrethroids have not been reported to provide effective control against *S. dorsalis* in the New World and although it causes an instant reduction in pest populations in other parts of the world, it also kills natural controlling agents, ultimately leading to resurgence of pest populations. Various formulations of imidacloprid (neonicotinoid insecticide class), used either as soil drenches or foliar applications provide effective suppression of *S. dorsalis* populations for several days (Table 3) after application of treatments.

Common Name	Trade Name	IRAC Class	Residual Control (days)			
			Foliar		Soil	
			Adult	Larva	Adult	Larva
Abamectin	Agrimek*, Avid*	6	2	2		
Acephate	Orthene*	1B	7	7	7	7
Chlorfenapyr	Pylon*	13	7	7	-	-
Dinotefuran	Venom*, Safari® SG	4A	10	15	0	0
imidacloprid	Marathon*, Provado*, Admire*	4A	15	15	15	15
Novaluron	Pedestal*, Ramon*	15	7-14	7-14	-	-
Spinosad	Conserve*, SpinTor*	5	15	15	_	_
Spinetoram	Radiant*	5	15	15	15	15
Thiamethoxam	Actara*, Platinum*	4A	10	15	10	15
Borax + orange oil + detergents	TriCon*	8D	10	10	-	-
Beauveria bassiana	Botanigard*	Not applicable	3-7	3-7	-	-
Metarhizium anisopliae	Met52*	Not applicable	7	7	-	-

Table 3. Choices of insecticides for rotational use against *Scirtothrips dorsalis* populations. Source: [80]

Management practices from an ecological point-of-view must be environmental friendly but from a growers' viewpoint must be economical, fast acting as well as long lasting. Different chemical insecticides that could satisfy all concerns, like spinetoram and various neonicotinoid insecticides do cause significant reduction in *S. dorsalis* on pepper crops [79]. However, due to their frequent use, insect pests are under intense selection pressure to develop resistance against these insecticides. There are many reports where excessive reliance on insecticides has resulted in resistance development in this pest. In India, resistance in *S. dorsalis* populations has been reported to a range of organochlorine (DDT, BHC and endosulfan), organophosphate (acephate, dimethoate, phosalone, methyl-o-demeton and triazophos) and carbamate (carbaryl) insecticides [82]. Recently, *S. dorsalis* was reported to develop resistance against monocrotophos, acephate, dimethoate, phosalone, carbaryl and triazophos [83]. Thus, in order to prevent or delay development of resistance or minimize the progressive assembly of genes for resistance through selection in the pest against a particular chemistry, it is necessary to rotate insecticides from diverse chemical groups, and explore alternative methods of pest control. Inclusion of effective biorational and biological products in a best management program for *S. dorsalis* can lead to reduced applications of synthetic insecticides. Use of biorational and biocontrol products early in the season will delay the buildup of damaging pest populations on host plants. Furthermore, reduction in the use of harmful insecticides will increase the population of natural biocontrol agents.

9.4. Biological control

Biological control is the active manipulation of beneficial organisms to reduce the pest population below the economic injury level [84]. In this, activity of one species is exploited to reduce adverse effects of another. It is one of the oldest types of pest management strategies. Biological control is employed with the aim of long time pest control by bringing the pest population to non-economic levels. Biological controlling agents are living natural enemies e.g. predators, parasitoid, parasites or pathogens. Various biological controlling agents like minute pirate bugs, *Orius spp.* (Hemiptera: Anthocoridae) and the phytoseiid mites *Neoseiulus cucumeris* and *Amblyseius swirskii* have been reported to provide effective control of *S. dorsalis* on pepper [85, 86]. Adults of *Orius insidiosus* have been observed to feed on all the developmental stages of thrips, and since it is a generalist predator which feeds on aphids, mites, moth eggs and pollen, its population does not decline when there are periodic drops in the thrips population. The biocontrol potential of two phytoseiid mites, *Neoseiulus cucumeris* and *Amblyseius swirskii* evaluated against *S. dorsalis* showed that *A. swirskii* can be a promising tool in managing its population on pepper [86]. In Japan, the predatory mite *Euseius sojaensis* was found to be effective in regulating *S. dorsalis* populations on grapes [87]. Other predatory phytoseiid mites that show promise for biological control include *E. hibisci* and *E. tularensis*. It has been suggested to use two or more natural enemies as a strategy to improve biological control of greenhouse pests [88]. Predators that warrant further study as potential natural enemies of *S. dorsalis* include lacewings (*Chrysoperla* spp.), several mirid bugs, ladybird beetles, and a number of predatory thrips including the black hunter thrips (*Leptothrips mali*), *Franklinothrips* (*Franklinothrips vespiformis*), the six spotted thrips (*Scolothrips sexmaculatus*), and the banded wing thrips (*Aeolothrips* spp.).

The role of entomopathogens like *Beauveria bassiana*, *Metarhizium anisopliae* and *Isaria fumosorosea* in managing field populations of *S. dorsalis* are still under study. *B. bassiana* used with some adjuvants has been reported to control larval populations of *S. dorsalis* for the first few days after application, but soon the population of *S. dorsalis* increases and becomes equivalent to the control plants [80]. In India, significant reduction in *S. dorsalis* populations was reported using entomopathogens *Fusarium semitectum* in pepper fields [89]. However, commercialization and success of this biorational product in different biogeographical regions is still in need of evaluation. Therefore, there is an immense need for developing new strategies to employ best management practices for this serious pest utilizing cultural, chemical and biological control methods.

10. Future prospects

Apart from changing climatic conditions, insect pests are another constraint affecting agricultural production. Insect pests are responsible for loss estimates of 10-20% of main agricultural crops which makes them a major yield limiting factor [90]. To control these pests chemical insecticides are often used by growers on a calendar basis which backfires many times and it leads to a "3R" situation - resistance, resurgence and replacement. To check this situation, it is important to utilize all the resources available in the agroecosystem in a controlled and effective manner. Integrated pest management is an ecosystem-based pest management strategy which focuses on the longtime control of pests using a combination of techniques, such as cultural control, biological control, habitat manipulation, and use of biotechnological methods. Chemical insecticides are used wisely only after monitoring, under suitable guidelines with the aim to control target pests with no effect on non-target organisms and environment. In the case of *S. dorsalis*, evaluation of chemical insecticides against effective predators like *A. swirskii*, *O. insidiosus* and *E. sojaensis* is needed so that both management systems can exist together.

In the near future, advancement in biological control strategies of *S. dorsalis* could be the use of banker plant systems. Our research group is working in this direction to screen and use different pepper cultivars which could be effectively used as banker plants for the establishment of predatory mites in nurseries and in field conditions. It can effectively solve a number of pest problems in ornamental and vegetable cropping systems including whiteflies, thrips and mites. Banker plant systems also known as open-rearing systems; it is an integrated biological control approach which involves combined aspects of augmentative and conservational biological control and habitat manipulation proposed as an efficient alternative to chemical based pest management techniques. [91, 92, 93]. Success of biological control strategies depends upon the potency of natural enemies against the target pest as well as its adaptability, survival and long-term establishment in the habitat. Installing banker plants in the agroecosystem, ornamental landscape and nurseries can support the establishment of biological control agents by providing suitable ecological infrastructures. The infrastructure can be in the form of a nutrient supplement (nectar/pollen) which is crucial for their survival in the absence of prey, or it can be in the form of a modified microhabitat which can provide protection against adverse abiotic conditions, an insecticide application as well as the hyper-predation/parasitism (secondary enemies) of the agents [94]. The provision of food and shelter reduces mortality of the released biological control agents and may favor their survival, fecundity, longevity and potency to regulate target pests in the habitat thereby supporting the success of biological control strategy.

Acknowledgements

This paper is submitted in partial fulfillment of the requirements for the PhD degree of the senior author. We express our heartful thanks to Dr. David Hall for allowing us to use SEM facility at USHRL, USDA-ARS in Fort Pierce, Florida. We appreciate Dr. Wayne Hunter for training in scanning electron microscope and John Prokop and Michael Cartwright for technical assistance during the experiment. Special thanks to Dr. David Schuster, Dr. Aaron Dickey for their constructive criticism and helpful comments which was important in the preparation of the previous version of this manuscript. Mention of any trade names or products does not imply endorsement or recommendation by the University of Florida or USDA.

Author details

Vivek Kumar[1,4]*, Garima Kakkar[2], Cindy L. McKenzie[3], Dakshina R. Seal[1] and
Lance S. Osborne[4]

*Address all correspondence to: vivekiari@ufl.edu

1 Tropical Research and Education Center, University of Florida, Institute of Food and Agricultural Sciences, Homestead, FL, USA

2 Fort Lauderdale Research and Education Center, University of Florida, Institute of Food and Agricultural Sciences, Davie, FL, USA

3 United States Department of Agriculture, Agricultural Research Services, Fort Pierce, FL, USA

4 Mid-Florida Research and Education Center, University of Florida, Institute of Food and Agricultural Sciences, Apopka, FL, USA

References

[1] Haseeb M, Kairo M, Flowers RW. New Approaches and possibilities for invasive pest identification using web-based tools. American Entomologist. 2011;57(4): 223-226.

[2] Chenje M, Mohamed-Katerere, J. Invasive alien species, In Mohamed-Katerere J. African Environment Outlook 2. UNEP: Nairobi, Kenya; 2006. p331-349.

[3] Pimentel D, Lach L, Zuniga R, Morrison D. Environmental and economic costs of nonindigenous species in the United States. Bioscience. 2000;50(1): 53-65.

[4] Reitz SR, Trumble JT. Competitive displacement among insects and arachnids. Annual Review of Entomology. 2002;47: 435-465.

[5] Simberloff D. Biological invasions: How are they affecting us, and what can we do about them? Western North American Naturalist. 2001;61(3): 308-315.

[6] Grimaldi DA, Shmakov A, Fraser N. Mesozoic thrips and early evolution of the order hysanoptera (Insecta). Journal of Paleontology. 2004;78(5): 941-952.

[7] Rugman-Jones PF, Hoddle MS, Mound LA, Stouthamer R. Molecular identification key for pest species of Scirtothrips (Thysanoptera: Thripidae). Journal of Economic Entomology. 2006;99(5): 1813-1819.

[8] Ananthakrishnan T. Bionomics of Thrips. Annual Review of Entomology. 1993;38: 71-92.

[9] EPPO. Scirtothrips dorsalis, In Smith IM, McNamara DG, Scott PR, Holderness M. [eds.], Quarantine Pests for Europe, 2nd Edition. CABI/EPPO, Wallingford; 1997. p14-25.

[10] Mound LA, Palmer JM. Identification, distribution and host plants of the pest species of Scirtothrips (Thysanoptera: Thripidae). Bulletin of Entomological Research. 1981;71(3): 467-479.

[11] Kumar V, Seal DR, Kakkar G, McKenzie CL, Osborne LS. New tropical fruit hosts of Scirtothrips dorsalis (Thysanoptera: Thripidae) and its relative abundance on them in south Florida. Florida Entomologist. 2012;95(1): 205-207.

[12] Seal DR, Klassen W, Kumar V. Biological parameters of Scirtothrips dorsalis (Thysanoptera:Thripidae) on selected hosts. Environmental Entomology. 2010;39(5): 1389-1398.

[13] Kumar V, Seal DR, Schuster DJ, McKenzie CL, Osborne LS, Maruniak J, Zhang S. Scirtothrips dorsalis (Thysanoptera: Thripidae): Scanning electron micrographs of key taxonomic traits and a preliminary morphometric analysis of the general morphology of populations of different continents. Florida Entomologist. 2011;94(4): 941-955.

[14] Kumar V. Characterizing phenotypic and genetic variations in the invasive chilli thrips, Scirtothrips dorsalis Hood (Thysanoptera: Thripidae). Doctoral dissertation. Entomology and Nematology department, University of Florida, Gainesville FL, USA. 2012.

[15] Hoddle MS, Heraty JM, Rugman-Jones PF, Mound LA, Stouthamer R. Relationships among species of Scirtothrips (Thysanoptera : Thripidae, Thripinae) using molecular and morphological data. Annals of Entomological Society of America. 2008;101(3): 491-500.

[16] Dev HN. Preliminary studies on the biology of Assam thrips, Scirtothrips dorsalis Hood on tea. Indian Journal of Entomology. 1964;26(00): 184-194.

[17] Ramakrishna Ayyar TV. Bionomics of some thrips injurious to cultivated plants in South India. Agriculture and Live-stock in India. 1932;2: 391-403.

[18] Ramakrishna Ayyar TV, Subbiah MS. The leaf curl disease of chilies caused by thrips in the Guntur and Madura tracks. The Madras Agricultural Journal. 1935;23: 403-410.

[19] Kulkarni GS. The "Murda" disease of chilli (Capsicum). Agricultural Journal of India. 1922;22: 51-54.

[20] MacLeod A, Collins D. CSL Report: Pest risk analysis for Scirtothrips dorsalis. Central Science Laboratory, Sand Hutton, York, UK; 2006.

[21] Pimentel D, Zuniga R, Morrison D. Update on the environmental and economic costs associated with alien-invasive species in the United States. Ecological Economics. 2005;52: 273-288.

[22] Ferriter A, Doren B, Goodyear C, Thayer D, Bruch J, Toth L, Bondle M, Lane J, Schmitz D, Pratt P, Snow S, Langeland K. The status of nonindigenous species in the south Florida Environment. South Florida environment report, South Florida Water management District, Florida Department of Environmental protection; 2006. p1-52.

[23] USDA. Port Information Network (PIN-309): Quarantine status database. US Department of Agriculture, Animal and Plant Health Inspection Service, Plant Protection and Quarantine, Riverdale, MD, USA; 2003.

[24] Silagyi AJ, Dixon WN. Assessment of chili thrips, Scirtothrips dorsalis Hood. Division of Plant Industry, Gainesville, Florida; 2006. p9.

[25] Holtz T. NPAG Report: Scirtothrips dorsalis Hood. New Pest Advisory Group, Center for Plant Health Science and technology, APHIS, USDA, Raleigh, North Carolina; 2006. p7.

[26] Ciomperlik MA, Seal DR. Surveys of St. Lucia and St. Vincent for Scirtothrips dorsalis (Hood), Jan. 14-23, 2004. USDA- APHIS PPQ, Technical report; 2004. p19.

[27] Hodges G, Edwards GB, Dixon W. Chilli thrips Scirtothrips dorsalis Hood (Thysanoptera: Thripidae) a new pest thrips for Florida. Florida Department of Agriculture and Consumer Service, Department of Plant Industries. 2005. On-line publication. http://www.freshfromflorida.com/pi/pest-alerts/scirtothrips-dorsalis.html (Accessed: March 14 2010).

[28] Venette RC, Davis EE. Chilli thrips/yellow thrips, Scirtothrips dorsalis Hood (Thysanoptera: Thripidae), Mini Pest Risk Assessment. University of Minnesota, St. Paul, MN, USA; 2004. p31.

[29] Meissner H, Lemay A, Borchert D, Nietschke B, Neeley A, Magarey R, Ciomperlik M, Brodel C, Dobbs T. Evaluation of possible pathways of introduction for Scirtothrips dorsalis Hood (Thysanoptera: Thripidae) from the Caribbean into the continental United States. Plant Epidemiology and Risk Assessment Laboratory, APHIS, USDA, Raleigh, North Carolina; 2005. p125.

[30] Morse JG, Hoddle MS. Invasion biology of thrips. Annual Review of Entomology. 2006;51: 67-89.

[31] Government of India, Ministry of Agriculture. Post harvest profile of chilli. Technical report; 2010. p1-80.

[32] Hosamani A. Management of chilli murda complex in irrigated ecosystem. Doctoral dissertation. University of Agricultural Sciences, Dharwad. India. 2007.

[33] Patel BH, Koshiya DJ, Korat DM. Population dynamics of chilli thrips, Scirtothrips dorsalis Hood in relation to weather parameters. Karnataka Journal of Agricultural Sciences. 2009;22(1): 108-110.

[34] Evans G. Recent trends in world and U.S. Mango production, trade and consumption. Food and Resource Economics Department, Florida Cooperative Extension Service, Institute of Food and Agricultural Sciences, University of Florida, Gainesville, FL; 2008. p1-7. On-line publication. http://edis.ifas.ufl.edu/pdffiles/FE/FE71800.pdf (Accessed: September 16 2012).

[35] Kwee L T, Chong KK. Diseases and disorders of mango in Malaysia. Art Printing Works SDN.BHD, Kuala Lumpur, Malaysia; 1994. p101.

[36] Aliakbarpour H, Che Salmah MR, Dieng H. Species composition and population dynamics of thrips (Thysanoptera) in mango orchards of northern peninsular Malaysia. Environmental Entomology. 2010;39(5): 1409-1419.

[37] Shibao M, Ehara S, Hosomi A, Tanaka H. Seasonal fluctuation in population density of phytoseiid mites and the yellow tea thrips, Scirtothrips dorsalis Hood (Thysanoptera: Thripidae) on grape, and predation of the thrips by Euseius sojaensis (Ehara) (Acari: Phytoseiidae). Applied Entomology and Zoology. 2004;39(4): 727-730.

[38] Masui S. Synchronism of immigration of adult yellow tea thrips, Scirtothrips dorsalis Hood (Thysanoptera: Thripidae) to citrus orchards with reference to their occurrence on surrounding host plants. Applied Entomology and Zoology. 2007;42(4): 517-523.

[39] Reitz SR, Yu-lin G, Zhong-ren L. Thrips: Pests of concern to China and the United States. Agricultural Sciences in China. 2011;10(6): 867-892.

[40] Garrett L. Summary table of projected economic losses following the possible establishment of the chilli thrips in the USA. USDA, APHIS report; 2004.

[41] ERS-USDA. Census of horticultural specialties report. 2011. Available at http://www.agcensus.usda.gov/Surveys/Census_of_Horticulture_Specialties/index.asp (Accessed: October 12 2011).

[42] Derksen A. Host susceptibility and population dynamics of Scirtothrips dorsalis Hood (Thysanoptera: Thripidae) on select ornamental hosts in southern Florida. Master thesis. Entomology and Nematology department, University of Florida, Gainesville FL, USA; 2009.

[43] FNGLA (Florida Nurserymen & Landscape Growers Association). The unlucky 13. Report of the Major Nursery Pest & Disease Identification Task Force. Florida Nursery Growers and Landscape Association, Orlando, Florida, United States; 2003. p1.

[44] Butani DK. Pests and diseases of chillies and their control. Pesticides. 1976;10: 38-41.

[45] Amin PW. Techniques for handling thrips as vectors of tomato spotted wilt virus and yellow spot virus of groundnut, Arachis hypogea L. Occasional Paper. Groundnut Entomology. ICRISAT. 1980;80: 1-20.

[46] Kodomari S. Control of yellow tea thrips, Scirtothrips dorsalis Hood, in tea field at east region in Shizuoka prefecture. Journal of Tea Research. 1978;48: 46-51.

[47] Global pest and disease database (GPDD). Report on GPDD Pest ID 1276 Scirtothrips dorsalis - APHIS; 2011. p1-15.

[48] Tatara A, Furuhashi K. Analytical study on damage to satsuma mandarin fruit by Scirtothrips dorsalis Hood (Thysanoptera: Thripidae), with particular reference to pest density. Applied Entomology and Zoology. 1992;36(4): 217-223.

[49] Chang NT. Major pest thrips in Taiwan. In Parker B L, Skinner M, Lewis T. [eds.], Thrips Biology and Management. Plenum Press, New York, United States; 1995. p105-108.

[50] Sanap MM, Nawale RN. Chemical control of chilli thrips, Scirtothrips dorsalis. Vegetable Science. 1987;14: 95-199.

[51] Amin BW. Leaf curl disease of chilli peppers in Maharashtra, India. Pans. 1979;25(2): 131-134.

[52] Seal DR, Ciomperlik MA, Richards ML, Klassen W. Distribution of chilli thrips, Scirtothrips dorsalis (Thysanoptera : Thripidae), in pepper fields and pepper plants on St. Vincent. Florida Entomologist. 2006;89(3): 311-320.

[53] EPPO. EPPO Standards - Diagnostic protocols for regulated pests - Scirtothrips aurantii, Scirtothrips citri, Scirtothrips dorsalis. OEPP/EPPO Bulletin. 2005;35: 353-356.

[54] Amin PW, Reddy DVR, Ghanekar AM. Transmission of tomato spotted wilt virus, the causal agent of bud necrosis of peanut, by Scirtothrips dorsalis and Frankliniella schultzei. Plant Disease. 1981;65: 663-665.

[55] Satyanarayana T, Reddy KL, Ratna AS, Deom CM, Gowda S, Reddy DVR. Peanut yellow spot virus: a distinct tospovirus species based on serology and nucleic acid hybridization. Annals of Applied Biology. 1996;129(2): 237-245.

[56] Rao P, Reddy AS, Reddy SV, Thirumala-Devi K, Chander RS, Kumar VM, Subramaniam K, Reddy TY, Nigam SN, Reddy DVR. The host range of tobacco streak virus in India and transmission by thrips. Annals of Applied Biology. 2003;142: 365-368.

[57] Chiemsombat P, Gajanandana O, Warin N, Hongprayoon R, Bhunchoth A, Pongsapich P. Biological and molecular characterization of tospoviruses in Thailand. Archives of Virolology. 2008;153: 571-577.

[58] Rugman-Jones PF, Hoddle MS, Stouthamer R. Nuclear-mitochondrial barcoding exposes the global pest western flower thrips (Thysanoptera: Thripidae) as two sympa-

tric cryptic species in its native California. Journal of Economic Entomology. 2010;103(3): 877-886.

[59] Vierbergen GB, Kucharczyk H, Kirk WDJ. A key to the second instar larvae of the Thripidae of the Western Palaearctic region (Thysanoptera). Tijdschrift voor Entomologische. 2010;153: 99-160.

[60] Skarlinsky TL. Identification aid for Scirtothrips dorsalis, Hood. USDA; 2004. http://mrec.ifas.ufl.edu/lso/DOCUMENTS/identification%20aid.pdf (Accessed: August 14 2011).

[61] Hoddle MS, Mound LA, Paris DL. Scirtothrips dorsalis. Thrips of California. University of California, California. USA; 2009. http://keys.lucidcentral.org/keys/v3/thrips_of_california/data/key/thysanoptera/Media/Html/browse_species/Scirtothrips_dorsalis.htm (Accessed: March 14 2011).

[62] Brunner PC, Fleming C, Frey JE. A molecular identification key for economically important thrips species (Thysanoptera: Thripidae) using direct sequencing and a PCRRFLP-based approach. Agricultural and Forest Entomology. 2002;4(2): 127-136.

[63] Rosen D. The role of taxonomy in effective biological control programs. Agriculture Ecosystems & Environment. 1986;15(2-3): 121-129.

[64] Asokan R, Kumar K, Kumar NKV, Ranganath HR. Molecular differences in the mitochondrial cytochrome oxidase I (mtCO1) gene and development of a species-specific marker for onion thrips, Thrips tabaci Lindeman, and melon thrips, T. palmi Karny (Thysanoptera: Thripidae), vectors of tospoviruses (Bunyaviridae). Bulletin of Entomological Research. 2007;97(5): 461-470.

[65] Farris RE, Ruiz-Arce R, Ciomperlik M, Vasquez JD, Deleon R. Development of a ribosomal DNA ITS2 marker for the identification of the thrips, Scirtothrips dorsalis. Journal of Insect Science. 2010;10: 1-15.

[66] Porco D, Rougerie R, Deharveng L, Hebert P. Coupling non-destructive DNA extraction and voucher retrieval for small soft-bodied arthropods in a high-throughput context: the example of Collembola. Molecular Ecology Resources. 2010;10(6): 942-945.

[67] Hunter SJ, Goodall TI, Walsh KA, Owen R, Day JC. Nondestructive DNA extraction from blackflies (Diptera: Simuliidae): retaining voucher specimens for DNA barcoding projects. Molecular Ecology Resources. 2008;8(1): 56-61.

[68] Funderburk J. Ecology of thrips, In Marullo R. and Mound L. (eds.) Thrips and tospoviruses: Proceedings of the 7th International Symposium on Thysanoptera 2001. CSIRO Entomology, Reggio Calabria, Italy; 2002. p-121-128.

[69] Nietschke BS, Borchert DM, Magarey RD, Ciomperlik MA. Climatological potential for Scirtothrips dorsalis (Thysanoptera: Thripidae) establishment in the United States. Florida Entomologist. 2008;91(1): 79-86.

[70] Shibao M, Tanaka F, Tsukuda R, Fujisaki K. Overwintering sites and stages of the chilli thrips, Scirtothrips dorsalis Hood (Thysanoptera: Thripidae) in grape fields. Japanese Journal of Applied Entomology and Zoology. 1991;35: 161-163.

[71] Masui S. Timing and distance of dispersal by flight of adult yellow tea thrips, Scirtothrips dorsalis Hood (Thysanoptera: Thripidae). Japanese Journal of Applied Entomology and Zoology. 2007;51: 137-140.

[72] Chu CC, Ciomperlik MA, Chang NT, Richards ML, Henneberry TJ. Developing and evaluating traps for monitoring Scirtothrips dorsalis (Thysanoptera: Thripidae). Florida Entomologist. 2006;89(1): 47-55.

[73] Tsuchiya M, Masui S, Kuboyama N. Color attraction of yellow tea thrips (Scirtothrips dorsalis Hood). Applied Entomology and Zoology. 1995;39(4): 299-303.

[74] Tsuchiya M, Furunishi K, Masui S. Behavior of yellow tea thrips (Scirtothrips dorsalis Hood) on a reflective sheet. Applied Entomology and Zoology. 1995;39(4): 289-297.

[75] Tsuchiya M, Masui S, Kuboyama N. Reduction of population denstiy of yellow tea thrips (Scirtothrips dorsalis Hood) on mandarin orange (Citrus unshiu Marc.) trees by application of white solution with/without reflective sheet mulching. Applied Entomology and Zoology. 1995;39(4): 305-312.

[76] George S, Giraddi RS, Patil RH. Utility of vermiwash for the management of thrips and mites on chilli (Capsicum annuum L.) amended with soil organics. Karnataka Journal of Agricultural Sciences. 2007;20(3): 657-659.

[77] Shibao M. Effects of insecticide application on population density of the chilli thrips, Scirtothrips dorsalis Hood (Thysanoptera: Thripidae), on grape. Applied Entomology and Zoology. 1997;32(3): 512-514.

[78] Asaf-Ali K, Abraham E, Thirumurthi S, Subramaniam T. Control of scabthrips (Scirtothrips dorsalis) infesting grapevine (Vitis vinifera). South Indian Horticulture. 1973;21(3): 113- 114.

[79] Seal DR, Ciomperlik MA, Richards ML, Klassen W. Comparative effectiveness of chemical insecticides against the chilli thrips, Scirtothrips dorsalis Hood (Thysanoptera: Thripidae), on pepper and their compatibility with natural enemies. Crop Protection. 2006;25(9): 949-955.

[80] Seal DR, Kumar V. Biological response of chilli thrips, Scirtothrips dorsalis Hood (Thysanoptera: Thripidae), to various regimes of chemical and biorational insecticides. Crop Protection. 2010;29: 1241-1247.

[81] Bethke J, Chamberlin J, Dobbs J, Faver M, Heinz K, Lindquist R, Ludwig S, McKenzie CL, Murphy G, Oetting R, Osborne L, Palmer C, Parrella M, Rechcigl N, Yates R. Thrips management program for plants for planting; 2010. Online publication. http://

mrec.ifas.ufl.edu/lso/DOCUMENTS/ThripsManagementProgram_100308.pdf.   (Accessed: August 29 2011).

[82] Reddy DVR, Ratna AS, Sudarshana MR, Poul R, Kumar IK. Serological relationships and purification of bud necrosis virus, a Tospovirus occurring in peanut (Arachis hypogaea L.) in India. Annals of Applied Biology 1992;120(2): 279-286.

[83] Vanisree PR, Upendhar S, Ramachandra Rao G, Srinivasa Rao V. Insecticide resistance in chilli thrips, Scirtothrips dorsalis (Hood) in Andhra Pradesh. In Harrison B. Resistant pest management newsletter No.2. Center for Integrated Plant Systems (CIPS); 2011. p22-25.

[84] Frank JH. Glossary of expressions in biological control. Univ. Florida; 2000. Online publication: http://ipm.ifas.ufl.edu/resources/extension_resources/glossary/ gloss.shtml#B (Accessed May 12 2012).

[85] Dogramaci M, Arthurs SP, Chen J, McKenzie CL, Irrizary F, Osborne L. Management of chilli thrips Scirtothrips dorsalis (Thysanoptera: Thripidae) on peppers by Amblyseius swirskii (Acari: Phytoseiidae) and Orius insidiosus (Hemiptera: Anthocoridae). Biological Control. 2011;59(3): 340-347.

[86] Arthurs S, McKenzie CL, Chen J, Dogramaci M, Brennan M, Houben K, Osborne L. Evaluation of Neoseiulus cucumeris and Amblyseius swirskii (Acari: Phytoseiidae) as biological control agents of chilli thrips, Scirtothrips dorsalis (Thysanoptera: Thripidae) on pepper. Biological Control. 2009;49(1): 91-96.

[87] Shibao M, Ehara S, Hosomi A, Tanaka H. Seasonal fluctuation in population density of phytoseiid mites and the yellow tea thrips, Scirtothrips dorsalis Hood (Thysanoptera: Thripidae) on grape, and predation of the thrips by Euseius sojaensis (Ehara) (Acari: Phytoseiidae). Japanese Journal of Applied Entomology and Zoology. 2004;39: 727-730.

[88] Chow A, Chau A, Heinz KM. Compatibility of Orius insidiosus (Hemiptera: Anthocoridae) with Amblyseius (Iphiseius) degenerans (Acari: Phytoseiidae) for control of Frankliniella occidentalis (Thysanoptera: Thripidae) on greenhouse roses. Biological Control. 2008;44(2): 259-270.

[89] Mikunthan G, Manjunatha M. Impact of habitat manipulation on mycopathogen, Fusarium semitectum to control Scirtothrips dorsalis and Polyphagotarsonemus latus of chilli. BioControl. 2008;53(2): 403-412.

[90] Ferry N, Edwards MG, Gatehouse JA, Gatehouse AMR. Plant-insect interactions: molecular approaches to insect resistance. Current Opinion in Biotechnology. 2004;15(2): 155-161.

[91] Osborne LS, Barrett J. You can bank on it; Banker plants can be used to rear natural enemies to help control greenhouse pests. Ornamental Outlook, Sept. 2005; p26-27.

Online publication: http://mrec.ifas.ufl.edu/lso/banker/Documents/BANKERFoliage.
pdf (Accessed: August 29 2011).

[92] Frank SD. Biological control of arthropod pests using banker plants systems: past
progress and future directions. Biological Control. 2010;52(1): 8-16.

[93] Xiao YF, Chen J, Cantliffe D, McKenzie CL, Houben K, Osborne LS. Establishment of
papaya banker plant system for parasitoid, Encarsia sophia (Hymenoptera: Aphili-
dae) against Bemisia tabaci (Hemiptera: Aleyrodidae) in greenhouse tomato production. Biological Control. 2011;58(3): 239-247.

[94] Landis DA, Wratten SD, Gurr GM. Habitat management to conserve natural enemies
of arthropod pests in agriculture. Annual Review of Entomology. 2000;45: 175-201.

Chapter 2

Companion Planting and Insect Pest Control

Joyce E. Parker, William E. Snyder,
George C. Hamilton and Cesar Rodriguez-Saona

Additional information is available at the end of the chapter

1. Introduction

There is growing public concern about pesticides' non-target effects on humans and other organisms, and many pests have evolved resistance to some of the most commonly-used pesticides. Together, these factors have led to increasing interest in non-chemical, ecologically-sound ways to manage pests [1]. One pest-management alternative is the diversification of agricultural fields by establishing "polycultures" that include one or more different crop varieties or species within the same field, to more-closely match the higher species richness typical of natural systems [2, 3]. After all, destructive, explosive herbivore outbreaks typical of agricultural monocultures are rarely seen in highly-diverse unmanaged communities.

There are several reasons that diverse plantings might experience fewer pest problems. First, it can be more difficult for specialized herbivores to "find" their host plant against a background of one or more non-host species [4]. Second, diverse plantings may provide a broader base of resources for natural enemies to exploit, both in terms of non-pest prey species and resources such as pollen and nectar provided by the plant themselves, building natural enemy communities and strengthening their impacts on pests [4]. Both host-hiding and encouragement of natural enemies have the potential to depress pest populations, reducing the need for pesticide applications and increasing crop yields [5, 6]. On the other hand, crop diversification can present management and economic challenges for farmers, making these schemes difficult to implement. For example, each of two or more crops in a field could require quite different management practices (e.g., planting, tillage and harvest all might need to occur at different times for the different crops), and growers must have access to profitable markets for all of the different crops grown together.

"Companion planting" is one specific type of polyculture, under which two plant species are grown together that are known, or believed, to synergistically improve one another's growth

[7]. That is, plants are brought together because they directly mask the specific chemical cues that one another's pests use to find their hosts, or because they hold and retain particularly effective natural enemies of one another's pests. In this chapter we define companion plants as interplantings of one crop (the companion) within another (the protection target), where the companion directly benefits the target through a specific known (or suspected) mechanism [8, 9]. Companion plants can control insect pests either directly, by discouraging pest establishment, and indirectly, by attracting natural enemies that then kill the pest. The ideal companion plant can be harvested, providing a direct economic return to the farmer [2] in addition to the indirect value in protecting the target crop. However, "sacrificial" companion plants which themselves provide no economic return can be useful when their economic benefit in increased yield of the target exceeds the cost of growing the companion [10, 11].

Companion planting has received less attention from researchers than other diversification schemes (such as insectary plants and cover crops), but this strategy is widely utilized by organic growers [8, 9]. Generally, recommendations on effective companion-target pairings come from popular press articles and gardening books, which make claims of the benefits of bringing together as companions aromatic herbs, certain flowers [12], or onions (*Allium* L. spp.) [13]; nearly always, vegetables are the protection target. However, these recommendations most-commonly reflect the gut-feeling experiences of particular farmers that these pairings are effective, rather than empirical data from replicated trials demonstrating that this hunch is correct. Indeed, more-rigorous examinations of companion-planting's effectiveness have yielded decidedly mixed evidence [e.g. 9, 14 and 15]. Here, we first review companion plants that disrupt host-location by the target's key pests, and then those that operate by attracting natural enemies of the protection target's pests. For companions operating through either mechanism, we discuss case-studies where underlying mechanisms have been examined within replicated field trials, highlighting evidence for why each companion-planting scheme succeeded or failed.

2. Companions that disrupt host location by pests

Herbivorous insects use a wide variety of means to differentiate between host and non-host plants. Consequently, host-finding behavior of the target's pests plays a key role in selecting an effective companion plant. Typically, host plant selection by insects is a catenary process involving sequences of behavioral acts influenced by many factors [16]. These can include the use of chemical cues, assessment of host plant size, and varying abilities to navigate and identify hosts among the surrounding vegetation. Therefore, both visual and chemical stimuli play key roles in host plant location and eventual acceptance. At longer distances, host-location often is primarily through the detection and tracking of a chemical plume [17]. At this scale, abiotic factors may play a strong role. For example, an odorous plume can be influenced not just by plant patch size, but also by temperature and wind speed, which can change the plume's spatial distribution and concentration [17]. As the insect draws near to the host plant, visual cues can increase in importance [17]. Visual indications that a suitable host has been located can include the size, shape and color of the plant [18]. Therefore, based on the dual roles of chemical and visual cues in host-location by herbivores, to be effective disruptors of host-location by the target's pests, companion plants would need to: (1) disrupt the ability of the pest to detect or recognize the target's chemical plume; (2) disrupt or obscure the visual profile of the target; or (3) act simultaneously through both chemical and visual disruption of host location.

Furthermore, ecological differences among pest species are likely to impact the effectiveness of companion planting. For example, specialist herbivores appear to be relatively strongly dissuaded from staying in diverse plantings where their host is just one component of the plant community, whereas generalist herbivores sometimes prefer diverse to simple plantings [19, 20]. Presumably this is because diverse plantings provide relatively few acceptable hosts per unit area for a specialist, but (potentially) several different hosts acceptable to a generalist. Likewise, the size/mobility of the pest is likely to be important. Potting et al. (2005) in reference [21] suggested that smaller sized arthropods such as mites, thrips, aphids and whiteflies that can be passively transported by wind currents, have limited host detection ability. Of course, when a pest moves haphazardly through the environment there is no active host-location behavior for a companion plant to disrupt! Apparently because insects that travel passively with wind currents may cause them to bypass trap crops leading to companion plant failure. Conversely, larger sized insects capable of direct flight have good sensory abilities that allow them to perform oriented movement and thus represent good candidates for control by companion planting [21].

2.1. Companions that draw pests away from the protection target

Trap crops are stands of plants grown that attract pest insects away from the target crop [11, 22] (Fig. 1).

Once pests are concentrated in the trap crop the pests can be removed by different means, such as burning or tilling-under the trap crop [11] or by making insecticide applications to the trap. A highly-effective trap crop can bring a relatively large number of pests into a relatively small area, such that pest management within the trap crop requires coverage of less ground than if the entire planting of the protection target had to be treated. Even if left unmanaged through other means, pests feeding within the trap are not damaging the protection target. Because trap crops are more attractive to the pest, they are usually rendered unmarketable due to pest damage. This means that, to be economically-viable, the cost of establishing and maintaining the trap crop must equal or exceed the value of crop-protection within the protection target.

There are many successful examples of trap cropping. For example, in California the need to spray for *Lygus* Hahn in cotton was almost completely eliminated due to the success of alfalfa trap crops [23-25]. In soybeans, Mexican bean beetles can be controlled using a trap crop of snap beans [26]. Similarly, for over 50 years in Belorussia early-planted potato trap crops have been used to protect later plantings of potatoes from Colorado potato beetle attack [27]. Even though many successful examples of trap crops have been reported, several studies have also demonstrated contradictory results with many declaring unsuccessful [28-31] to unreliable control of pests [32]. For example, Luther et al. (1996) in reference [29] explored trap crops of Indian mustard and Tastie cabbage to control diamondback moth and *Pieris rapae* L. in Scorpio

Figure 1. A trap crop of Pacific gold mustard (companion plant) is flanked on both sides by broccoli (target crop). The symbols (+) represent the principal mechanism at work. Here, the trap crop, designated with two (+) signs, are more attractive than the protection target-broccoli. The mustard trap crop is used to attract pest insects away from broccoli.

cabbage and discovered that these trap crops were effective at attracting these pests; however, the distance between the trap crop and the protection target allowed for pests to spillover back into the protection target. In another experiment using Indian mustard as a trap crop, Bender et al. (1999) in reference [30] intercropped Indian mustard with cabbage to control lepidopterous insects and found that Indian mustard did not appear to preferentially attract insect pests. Overall, the relative effectiveness of the trap crop depends on the spatial dimensions of the trap crop and protection target, the trap crop and protection target species and pest behavior.

The need to control pests in the trap crop can be avoided when "dead-end" traps are deployed. Dead-end trap cropping utilizes specific plants that are highly preferred as ovipositional sites, but incapable of supporting development of pest offspring [33, 34]. For example, the diamond back moth (*Plutella xylostella* L.), a pest of *Brassica* crops, is highly attracted for oviposition to the G-type of yellow rocket (*Barbarea vulagaris* R. Br.), but the larvae are not able to survive on this host plant [35]. This inability to survive has been attributed to a feeding deterrent, monodesmosidic triterpenoid saponin [36] and so larvae cannot complete development.

Similarly, potato plants genetically engineered to express *Bacillus thuringiensis* (Bt) proteins that are deadly to the Colorado potato beetle, and planted early in the season, can act as dead-end traps that kill early-arriving potato beetles [37].

Trap crop effectiveness can be enhanced by incorporating multiple plant species simultaneously. Diverse trap crops include plants with different chemical profiles, physical structures and plant phenologies, therefore, diverse trap crops may provide for a more attractive trap crop. For example, in Finland, mixtures of Chinese cabbage, marigolds, rape and sunflower were used successfully as a diverse trap crop to manage the pollen beetle (*Melighetes aeneus* F.) in cauliflower [38]. Furthermore, Parker (2012) in reference [39] conducted experiments exploring the use of simple and diverse trap crops to control the crucifer flea beetle (*Phyllotreta cruciferae* Goeze) in broccoli (*Brassica oleracea* L. var. *italica*). The trap crops included mono-cultures and polycultures of two or three species of Pacific gold mustard (*Brassica juncea* L.), pac choi (*Brassica rapa* L. subsp. *pekinensis*) and rape (*Brassica napus* L.). Results indicated that broccoli planted adjacent to diverse trap crops containing all three trap crop species attained the greatest dry weight suggesting that the trap crops species were not particularly effective when planted alone, however, provided substantial plant protection when planted in multi-species polycultures. Thus, diverse trap crops consisting of all three trap crop species (Pacific gold mustard, pac choi and rape) provided the most effective trap crop mixture.

The success of trap crops depends on a number of variables, such as the physical layout of the trap crop (e.g., size, shape, location) and the pests' patterns of movement behavior [40]. For example perimeter trap crops, trap crops sown around the border of the main crop [41], have been used to disrupt Colorado potato beetle (*Leptinotarsa decemlineata* Say) colonization of potato fields from overwintering sites that ring the field [42-44]. However, depending on the pest targeted for control and the cropping system, perimeter trap crops may not be the most effective physical design. For example, a perimeter trap crop may not impede pest movement if the pest descends on a crop from high elevations. In reference [11] Hokkanen (1991) has recommended an area of about 10% of the main crop area be devoted to the trap. A smaller trap crop planting leaves more farm ground available for planting marketable crops.

In general, throughout trap cropping literature, trap crops are most effective when they are attractive over a longer period of time than the target crop, and when trap crops target mobile pests that can easily move among the trap and protection-target plantings [11]. References [11] and [41] reported trap crop success particularly with larger beetles [11] and tephritid flies [41], insects generally capable of direct flight.

2.2. Plants that repel

Plants with aromatic qualities contain volatile oils that may interfere with host plant location, feeding, distribution and mating, resulting in decreased pest abundance [45-47] (Fig. 2).

Moreover, certain plants contain chemical properties which can repel or deter pest insects and many of these products are used to produce botanical insecticides. For example, pyrethrum obtained from dried flower of the pyrethrum daisy (*Tanacetum cinerariaefolium* L.), neem extracted from seeds of the Indian neem (*Azadirachta indica* A. Juss.) and essential oils extracted

Figure 2. Intercroppings of spring onions (companion plant) are implemented to protect broccoli (target crop) from pest attack. Here, spring onions are used as a repellent to push pest insects away from broccoli. The symbols represent their potential attractive (+) and repellent (-) properties.

from herbs such as rosemary, eucalyptus, clove, thyme and mint have been used for pest control [48]. Generally, aromatic herbs and certain plants are recommended for their supposed repellent qualities. For example, herbs such as basil (*Ocimum basilicum* L.) planted with tomatoes have been recorded to repel thrips [49] and tomato hornworms [50]. Plants in the genus *Allium* (onion) have been observed to exhibit repellent properties against a variety of insects and other arthropods including moths [51], cockroaches [52], mites [53] and aphids [54]. These examples represent a wide array of arthropods that respond to repellent odors and demonstrate the potential repellent plant properties can have on pest control.

Furthermore, many studies have reported a wide variety of companion plants to contain repellent properties against pests of *Brassica* crops. *Brassica* species are an economically important crop throughout the world [55], sometimes comprising up to 25% of the land devoted to vegetable crops [56]. These companion plants included sage (*Salvia officinalis* L.), rosemary (*Rosemarinus officinalis* L.), hyssop (*Hyssop officinalis* L.), thyme (*Thymus vulgaris* L.), dill (*Anethum graveolens* L.), southernwood (*Artemisia abrotanum* L.), mint (*Menta* L. spp.), tansy (*Tanacetum vulgare* L.), chamomile (several genera), orange nasturtium (*Tropaeolum Majus* L.)

[57], celery and tomatoes [57, 58]. Similarly, intercropping tomatoes with cabbages has been suggested to repel the diamond back moth [59] and ragweed (*Ambrosia artemisifolia* L.) has been used to repel the crucifer flea beetle (*Phyllotreta cruciferae*) from collards (*Brassica olera-cea* L. var. *acephala*) [60], both widespread pests of *Brassica* crops.

Not all studies using repellent companion plants have reported positive results. Early data have suggested no scientific evidence that odors from aromatic plants can repel or deter pest insects [61]. In reference [62] Latheef and Irwin (1979) found no significant differences in the number of eggs, larvae, pupae, or damage by cabbage pests between companion plants; French marigold (*Tagetes patula* L.), garden nasturtium pennyroyal (*Mentha pulegium* L.), peppermint (*Mentha piperita* L.), garden sage, thyme and control treatments. Furthermore, French mari-golds (*Tagetes patula* L.) intercropped in carrots did not repel the carrot fly (*Psila rosae* F.) [47]. Even reports of frequently recommended companion herbs did not always improve pest control. For example, there were no differences in diamond back moth oviposition between Brussels sprouts (*B. oleracea*) intercropped with sage (*S. officinalis*) and thyme (*T. vulgaris*) [61]. Sage and thyme represent two common companion plants noted for their pungent odors [9]. Billiald et al. (2005) in reference [63] and Couty et al. (2006) in reference [64] concluded that if these highly aromatic plants were truly repellent, insects would not land on non-host com-panion plants.

Indeed, other mechanisms other than repellent odors might have a prominent role in plant protection. In reference [61] Dover (1986) noted reduced oviposition by the diamond back moth caused by contact stimuli and not repellent volatiles of sage and thyme. Therefore, sage and thyme were still protecting the target crop; however, this protection was caused by alternative mechanisms other than repellent odors. Similarly, research has demonstrated that aromatic plants such as marigolds (*Tagetes erecta* L.) and mint (*Mentha piperita* L.) did not repel the onion fly (*Delia antiqua* Meigen) or the cabbage root fly (*D. radicum* L.), but instead disrupted their normal chain of host plant selecting behaviors [16, 65, 66].

The response to a repellent plant will vary depending on the behavior of the insect and the plant involved. As a result, a repellent plant that can be effective for one pest might not provide effective control for another [67]. Finally, many experiments to determine plant's repellent capabilities were carried out in laboratory settings and do not necessarily represent field conditions [9].

2.3. Plants that mask

Companion plants may release volatiles that mask host plant odors [59, 60, 68] interfering with host plant location (Fig. 3).

For example, host location by the cabbage root fly (*D. radicum*) was disrupted when host plants were surrounded by a wide variety of plants including weeds and marketable crops [69, 70] such as spurrey (*Spergula arvensis* L.) [71], peas (*Pisum sativum* L.) [72], rye-grass (*Lolium perenne* L.) [72] or clover [73, 74]. However, Finch and Collier (2000) in reference [9] suggested that even though these diverse companion plants contain different chemical profiles, it is unlikely that all would be able to mask host plant odors. Further research has demonstrated

Figure 3. Marigolds (companion plant) are intercropped with broccoli (target crop) to interfere with host plant loca-tion. Here, several mechanisms may be involved in protecting broccoli including masking host plant odors or visually camouflaging broccoli making it less apparent. Here, the symbol (+) is shaded to represent a less apparent target crop-broccoli.

that in a wind tunnel, cabbage root fly move toward *Brassica* plants surrounded by clover just as much as *Brassica* plants grown in bare soil indicating that odors from clover did not mask those of the *Brassica* plants [75].

In addition to hiding odors emitted by the protection target, companion plants have also been reported to alter the chemical profile of the protection target. For example, certain companion plants can directly affect adjacent plants by chemicals taken up through its roots [76]. African marigolds (*Tagetes* spp.) produce root exudates which can be absorbed by neighboring plants [77] and may help to explain the reports of African marigold reducing pest numbers [9]. African marigolds also release thiopene, which acts as a repellent to nematodes [78]. Similarly, studies exploring various barley cultivars discovered that airborne exposure of certain combinations of undamaged cultivars caused the receiving plant to become less acceptable to aphids [79-81] and this was also confirmed in field settings [80]. Thus, volatile interactions between odors of host and non-host plants and even single species with different cultivars can affect the behavior of pest insects.

2.4. Plants that camouflage or physically block

In addition to protecting crops with olfactory cues, companion plants may also physically and visually camouflage or block host plants [9, 14, 15, 20, 47, 60, 82]. The 'appropriate/inappropriate landing' theory proposed that green surfaces surrounding host plants may disrupt host plant finding [9]. The 'appropriate/inappropriate landing' theory was originally inspired from studies exploring the oviposition behavior of cabbage herbivores and found that reduced damage in intercropping systems were attributed to a disruption of oviposition behavior [9]. This can occur when insects land on a companion plant instead of the target crop before or during oviposition [83]. For example, Atsatt and O'Dowd (1976) in reference [84] demonstrated that *Delia radicum* (L.) (cabbage root fly) spent twice as much time on a non-host plant after landing on it compared to a host plant. This demonstrated that companion plants can disrupt and arrest *D. radicum* on inappropriate hosts (companion plants). Consequently, *D. radicum* will start its oviposition process from the beginning which may reduce the total number of eggs layed on the target crop. Studies have found similar post-alighting behavior of *Delia floralis* Fallén (turnip root fly) and the decision to oviposit after landing on host and non-host plants [85, 86].

Companion plants may visually (Fig. 3) or physically (Fig. 4) obstruct host plant location rendering host plants less apparent [87].

For example, host plant location in the crucifer flea beetle (*Phyllotreta cruciferae* Goeze) is disrupted when non-host plant foliage, either visual or hidden, is present [60]. Similarly, Kostal and Finch (1994) in reference [72] and Ryan et al. (1980) in reference [88] both showed that artificial plant replicas made from green card or green paper could disrupt host plant location. Companion plant height is also an important factor in pest suppression. Tall plants can impede pest movement within a cropping system [89]. For example, maize has been used to protect bean plants from pest attack [90] and dill has been used as a vegetative barrier to inhibit pest movement in organic farms (personal observation). Frequently recommended companion plants used as physical barriers include sunflowers, sorghum, sesame and peal-millet [91]. In addition, companion plant barriers may also be used to reduce the spread and transmission of insect vectored viruses [92].

Nevertheless, these mechanisms may not rely solely on physical obstruction [93]. For example, the presence of desiccated clover plants (brown in color), which retained the same architecture as living plants (green in color), but only differed in their appearance from living plants, did not reduce the number of cabbage root fly (*D. radicum*), diamond back moth (*P. xylostella*) and the large white butterfly (*Pieris brassicae* L.) eggs when compared to the target crop on bare ground [93]. However, when live clover surrounded the target crop, the numbers of eggs laid were reduced suggesting that the physical presence of clover alone was not enough to prevent a reduction in oviposition [93]. Therefore, the size, shape, color and chemical profiles of companion plants may interact together reducing pest numbers making it is difficult to tease apart specific mechanisms which may be contributing to pest control.

Figure 4. Dill (companion plant) is used as a physical barrier to protect broccoli (target crop) from pest attack. Here, the height of the dill can impede pest movement.

2.5. Combinations of companion planting techniques

In some systems, different companion planting methods have been combined to work synergistically and improve pest control. For example, in Kenya trap crops have been combined with repellent plants and implemented successfully in a 'push-pull' system [94] to control spotted stem borer (*Chilo partellus* Swinhoe) in maize (*Zea mays* L.) [95, 96]. The repellent plants included a variety of non-host plants such as molasses grass (*Melinis minutiflora* P. Beauv.), silverleaf desmodium (*Desmodium uncinatum* Jacq.) or green leaf desmodium (*Desmodium intortum* Mill.) and the trap crop plantings included Napier grass (*Pennisetum purpurerum* Schumach) or Sudan grass (*Sorghum vulgare sudanense* Hitchc.) [94]. Here, the 'push' (repellent companion plants) drives the pest insect away from the target crop while the 'pull' (trap crop) simultaneously lures the pests toward the trap crop. Kahn and Pickett (2003) in reference [96] have reported thousands of farmers in east Africa to utilize the push-pull strategies to protect maize and sorghum. In addition, Komi et al. (2006) in reference [97] suggested that maize-legumes or maize-cassava intercrops can provide a 'push' for push-pull systems incorporating Jack-bean (*Canavalia ensiformis* L.) as a highly attractive trap crop 'pull'. The goal of the push-pull strategy aims to minimize negative environmental consequences and maximize pest control, sustainability and crop yield [94].

3. Plants that enhance conservation biological control

While the previous theories explored bottom-up forces in which companion plants improved pest control, Root (1973) 'enemies hypothesis' in reference [4] explored top-down mechanisms. He proposed that natural enemy populations are greater in polycultures because diverse habitats provide a greater variety of prey and host species that become available at different times. Furthermore, a greater diversity of prey and host species allows natural enemy populations to stabilize and persists without driving their host populations to extinction [4]. Altogether these theories present processes which may contribute to the lower abundance of pest insects in mixed cropping systems. Not surprisingly, companion plants may provide pest control by one or several of these mechanisms.

Pest populations can be managed by enhancing the performance of locally existing communities of natural enemies [98]. This can be accomplished by incorporating non-crop vegetation, such as flowering plants also known as insectary plants, into a cropping system (Fig. 5).

Figure 5. Flowering companion plants are incorporated into this mixed vegetable farm to enhance the efficacy of natural enemies and improve pest suppression.

Companion plants can provide essential components in conservation biological control by serving as an alternative food source and supplying shelter to natural enemies [99]. Many natural enemies including predators and parasitoids require non-prey food items in order to develop and reproduce [100-102]. For example, adult syrphids whose larvae are voracious predators of aphids, feed on both pollen and nectar [103]. Pollen and nectar are essential resources for natural enemies which satisfy different health requirements. Nectar is a source for carbohydrates and provides energy, while pollen supplies nutrients for egg production [103-106]. In wheat fields in England and in horticultural and pastoral habitats in New Zealand over 95% of gravid female syrphids were found with pollen in their gut [103]. As a result, flowering plants can increase the fecundity and longevity of parasitic hymenoptera [107-109] and predators [110, 111]. In addition to increasing natural enemy fitness, improved nutrition may also enhance foraging behavior [e.g. 112, 113] and increase the female-based sex ratio of parasitoid offsprings [114]. A wide variety of natural enemies utilize non-prey food sources. For example, pollen and nectar have been demonstrated to be highly attractive to variety of predators including syrphids [103, 115, 116], coccinellids [117-119], and lacewings [117].

One method to increase natural enemy density using companion plants includes incorporating certain flowering plants into a cropping system. This is often accomplished by planting flowering strips or border plantings in crop fields. Plants in the family Apiaceae are highly attractive to certain beneficial insect populations and are generally recommended as insectary plants [120]. This can be attributed to their exposed nectaries and the structure of their compound inflorescence which creates a "landing platform" [121, 122]. In addition, natural enemies are attracted to the field by the color and odor of companion plants [123]. Another commonly used insectary plant is *Phacelia tanacetifolia* Benth, which has been employed in borders of crop fields because it produces large amounts of pollen and nectar [124, 125]. For example, White et al. (1995) in reference [116] incorporated plantings of *P. tanacetifolia* near cabbage (*B. oleracea*) to increase syrphid densities to control aphids. Similarly, MacLeod (1992) in reference [126] and Lövei et al. (1993) in reference [127] demonstrated that syrphids are highly attracted to the floral resources provided by coriander and buckwheat. Companion plants may work simultaneously influencing both top-down and bottom-up mechanisms. For example, while some studies have demonstrated dill to improve pest control by containing repellent properties, other studies have indicated that dill may also increase predator populations. Patt et al. (1997) in reference [128] found reduced survivorship and populations of Colorado potato beetle (*L. decemlineata*) when dill was intercropped with eggplant and attributed the lower pest numbers to improved biological control.

Flowering companion plants have been used in different cropping systems to enhance the impact of natural enemies. For example, in organic vineyards, [110, 111] increased natural enemies by supplying access to nectar-producing plants such as alyssum (*Lobularia maritima* L.). Other various herbs have also been used this way in Europe [126, 129, 130] and in New Zealand [115, 127]. Overall, flowering companion plants have been implemented in a variety of crops including cereals, vegetable crops and fruit orchards [99, 131-137] to improve conservation biocontrol. In addition to food resources, companion plants can provide shelter from predators and pesticides as well as favorable microclimates [138, 139] including overwintering sites [140]. Furthermore, companion plants can also influence the spatial distribution of natural enemies in and around crops [141, 142] improving pest control.

Indeed, the advantages of plant-based resources for natural enemies have only recently been recognized by major reviews [99, 143- 146], and the growing empirical evidence has demonstrated their importance in pest suppression. However, the interactions between the companion plant, target pests and their natural enemies are complex. For example, incorporating companion plants may not necessarily improve biological control if the flowering does not coincide with the activity of natural enemies [147], or if natural enemies do not move from the companion plants to the target crop [117, 148]. Moreover, plant structures, such as the corolla, may obstruct feeding by natural enemies [128] and diverse habitats may complicate prey location by predators and parasitoids [143,149, 150]. Just as pest insects may react differently to the same companion plant, predators within the same family can also respond to similar companion plants in different ways. For example, certain syrphids are highly specialized feeders, while others are generalist [151] influencing companion plant selection. However, the possible obstructions to conservation biocontrol can be diminished. One way to improve the effectiveness of companion plants in conservation biocontrol is to select plants that benefit key natural enemies [152]. Again, this highlights the importance of implementing "careful diversification" as a pest management method [144, 153-155]. Overall, incorporating companion plants to enhance biological control holds promise for managing pests in crops.

Companion plants have also been used as banker plants. Banker plants are usually non-crop species that are deliberately infested with a non-pest insect and improves biological control by providing natural enemies with alternative prey [e.g. 156-158, but see 159, 160] even in the absence of pests [e.g. 156, 159, 161]. This allows natural enemy populations to reproduce and persists throughout the season. Banker plants have been used in both conservation and augmentative biological control programs. Many studies have used banker plants consisting of wheat or barley to sustain populations of the bird cherry-oat aphid (*Rhopalosiphum padi* L.) because this aphid feeds only on members of the Poaeceae family and does not pose a threat for vegetable and ornamental production [162]. However, success can be variable. Jacobson and Croft (1998) in reference [163] compared wheat, rye and corn as banker plants in its ability to sustain the bird cherry-oat aphid parasitoid (*Aphidius colemani* Viereck) and found that control was dependent on banker plant density, release rate and season. One successful example was implemented in apple orchards. To control the rosy apple aphid (*Dysaphis plantaginea* Passerini) in apple orchards, Bribosia et al. (2005) in reference [164] used Rowan trees (*Sorbus aucuparia* L.) as banker plants to maintain densities of Rowan aphids (*Dysaphis sorbi* L.) which served as an alternate host for the braconid parasitoid *Ephedrus persicae* Froggatt.

4. Constraints and challenges

Incorporating companion plants into pest management strategies is not without challenges. Farmers often face logistical constraints when incorporating companion plants into their field designs. For example, modern agriculture techniques and equipment are not conducive to growing multiple crops in one field [165]. Furthermore, companion plants may hinder crop yield and reduce economical benefits [166, 167]. Beizhou et al. (2011) in reference [168] reported an outbreak of secondary pests and reduced yield in an orchard setting. Decreased yields can often be attributed to competition for resources by incorporating inappropriate companion plants [169]. In certain cases, vegetational diversification can diminish the impacts of biological control. Generally, greater habitat diversity leads to a greater abundance of prey and host species. For instance, improved diversity can lead to reduced biological control by generalist predators which can be influenced by the greater diversity and abundance of alternative prey [123]. Straub et al. (2008) in reference [152] reviewed findings from natural enemy diversity experiments and found that results can range from negative (reduced control) to positive (improved control) due to effects from intraguild predation and species complementarity.

Therefore, choosing which type of companion plant to incorporate in a diversification scheme is challenging. For example, plant phenology, attractiveness and accessibility of the flowers to natural enemies [128] and pest species will play a key role in plant selection. However, it is possible to minimize the reductions in economic returns within companion planting schemes. It is important to use plants that can provide a satisfactory economic return, if possible, as compared to the target crop planted in monoculture [170]. In conservation biocontrol, to reduce negative impacts from biocontrol antagonists or the targeted pest, Straub et al. (2008) in reference [152] suggested using specific resources that can selectively benefit key natural enemies. Overall, whether companion plants control pests through bottom-up or top-down mechanism, their impact will depend on companion plant selection. This emphasizes the significance of finding the "right type" of diversity that combines species that complement one another in ecologically-relevant ways [67].

Designing companion planting schemes pose several impending issues. For instance, optimal distances between the companion plant and the target crop needs to be determined before specific recommendations can be made. The distance to which an insect is attracted to a source has proven to be variable and is a key area in companion plant success. Evans and Allen-Williams in reference [171] demonstrated that attraction can occur at distances of up to 20 m. Judd and Borden (1989) in reference [172] showed attraction of up to 100 m, however, other researchers have shown distances of only a few centimeters [173-176]. Therefore, adjusting the design depending on the insect's behavior and movement [83], the insect's search mode [177, 178] and diet breadth [20] may be necessary for companion plant success. Furthermore, an insect's feeding behavior will affect the success of companion plants in pest management strategies. For example plant structure can affect herbivory. Rape (*B. napus*) can be composed of trichomes that are nonglandular and simple or unbranched [179] and in some cases act as physical barriers that complicate feeding [180].

5. Conclusions

Many examples of companion plant to reduce pest numbers have been demonstrated; fewer diamondback moths were found on Brussels sprouts when intercropped with malting barley, sage or thyme [61]. Similarly, lower numbers of striped flea beetles were observed when Chinese cabbage (*Brassica chinensis* L.) was intercropped with green onions (*Allium fistulosum* L.) [181], while Mutiga et al. (2010) in reference [182] recorded significantly lower numbers of the cabbage aphid (*Brevicoryne brassicae* L.) when spring onion (*Allium cepa* L.) was intercropped with collard (*B. oleracea* var. *acephala*). However, the mechanisms through which companions protect the target are not well understood [183]. Many studies have suggested that chemical properties in the plant can repel insects [94], while others have suggested that companion plants are considered chemically neutral [66]. For example, Finch et al. (2003) in reference [66] demonstrated that commonly grown companion plants used for their repellent properties, marigolds and mint, did not repel the onion fly or the cabbage root fly (*D. radicum*), but rather interrupted their host finding and selecting behaviors [16, 65]. Thus, even though the companion plants did not repel pests, they were still able to disrupt host plant finding through alternative mechanisms. Overall, the effectiveness of companion plants to reduce pest numbers is not being disputed, but rather the mechanisms in which they work.

Caution should be taken before using companion plants in pest management as results can be mixed. For example, experiments conducted by Held et al. (2003) in reference [12] explored several putative companion plants in their ability to deter Japanese beetles (*Popillia japonica* Newman) from damaging roses and concluded that companion plants were unlikely to help. Diversifying cropping schemes is an essential step in the future of pest management. Companion planting represents just one of many areas in which a single farmer can incorporate diversifying schemes to reduce pest densities in an in-field approach. However, relatively subtle factors may determine whether crop-diversification schemes succeed or fail in improving pest suppression and crop response. Therefore, further research is needed on understanding the interactions between plant selection, mechanisms of benefit and patterns in time and crop phenology. Ultimately, cultural control strategies like companion planting can conserve species diversity, reduce pesticide use and enhance pest control.

Author details

Joyce E. Parker[1], William E. Snyder[2], George C. Hamilton[1] and Cesar Rodriguez-Saona[1]

1 Department of Entomology, Rutgers University, New Brunswick, NJ, USA

2 Department of Entomology, Washington State University, Pullman, WA, USA

References

[1] Denholm I, Devine G J and Williamson M S. Evolutionary genetics. Insecticide resistance on the move. Science 2002;297: 2222-2223.

[2] Vandermeer J H. The ecology of intercropping. Cambridge University Press, New York, New York, USA; 1989.

[3] Altieri M A and Nicholls C. Biodiversity and pest management in agroecosystems. New York, USA, Haworth Press; 1994.

[4] Root R B. Organization of a plant-arthropod association in simple and diverse habitats: fauna of collards (Brassica oleracea). Ecological Monographs 1973;43: 95-120.

[5] Pimentel D. Species diversity and insect population outbreaks. Annals of the Entomological Society of America 1961;54: 76-86.

[6] van Emden H F and Williams G F. Insect stability and diversity in agroecosystems. Annual Reviews of Entomology 1974;19: 455-475.

[7] Franck G. Companion planting: Successful gardening the organic way. Thorsons Pub. Group , Wellingborough, Northamptonshire, England; 1983.

[8] Cunningham S J. Great garden companions: a companion planting system for a beautiful, chemical-free vegetable garden. Rodale, Emmaus, PA; 1998.

[9] Finch S and Collier R H. Host-plant selection by insects – a theory based on 'appropriate/inappropriate landings' by pest insects of cruciferous plants. Entomologia Experimentalis 2000;96: 91-102.

[10] Altieri M A. The ecological role of biodiversity in agroecosystems. Agriculture, Ecosystem and Environment 1999;74: 19-31.

[11] Hokkanen H M T. Trap cropping in pest management. Annual Review of Entomology 1991;36: 119-138.

[12] Held D W, Gonsiska P and Potter D A. Evaluating companion planting and non-host masking odors for protecting roses from the Japanese Beetle (Coleoptera: Scarabaeidae). Journal of Economic Entomology 2003; 96: 81-87.

[13] Yepsen R B. The encyclopedia of natural insect and disease control. Rodale, Emmaus, PA; 1984.

[14] Thiery D and Visser J H. Masking of host plant odour in the olfactory orientation of the Colorado potato beetle. Entomologia Experimentalis et Applicata 1986;41: 165-172.

[15] Thiery D and Visser J H. Misleading the Colorado potato beetle with an odor blend. Journal of Chemical Ecology 1987;13: 1139-1146.

[16] Atkins M D. Introduction to Insect Behaviour. Macmillan, New York, NY, USA; 1980.

[17] Schoonhoven L, van Loon J J A and Dicke M. Insect-Plant Biology. Oxford University Press, New York; 2005.

[18] Prokopy R J and Owens E D. Visual detection of plants by herbivorous insects. Annual Review of Entomology 1983;28: 337-364.

[19] Bach C E. Effects of plant density and diversity on the population dynamics of a specialist herbivore, the striped cucumber beetle, Acalymma vitata (Fab.) Ecology 1980;61: 1515-1530.

[20] Andow D A. Vegetational diversity and arthropod population response. Annual Review of Entomology 1991;36: 561-586.

[21] Potting R, Perry J, and Powell JNW. Insect behavioural ecology and other factors affecting the control efficacy of agroecosystem diversification strategies. Ecological Modeling 2005;182: 199-216.

[22] Shelton A M and Badenes-Perez F R. Concepts and applications of trap cropping in pest management. Annual Review of Entomology 2006;51: 285-308.

[23] Stern V M, Mueller A, Sevacherian V and Way M. Lygus bug control in cotton through alfalfa interplanting. California Agriculture 1969;23: 8-10.

[24] Sevacherian V and Stern V M. Host plant preferences of Lygus bugs in alfalfa-interplanted cotton fields. Environmental Entomology 1974;3: 761-766.

[25] Stern V M. Environmental control of insects using trap crops, saniation, [prevention and harvesting. in Handbook of Pest Management in Agriculture. D. Pimentel (ed). Vols. 1,2. Boca Raton, FL; 1981.

[26] Rust R W. Evaluation of trap crop procedures for control of Mexican bean beetle in soybeans and lima beans. Journal of Economic Entomology 1977;70: 630-632.

[27] Dorozhkin N A, Bel'skaya S I and Meleshkevich A A. Combined protection of potatoes. Zasch. Rast 1975;7: 6-8.

[28] Silva-Krott I U, Singh P, Lali T S and Maniappan R. Development of a trap cropping system for cabbage in Guam. Pest Management Horticultural Ecosystems 1995;1: 27-35.

[29] Luther G C, Valenzuela H R and DeFrank J. Impact of cruciferous trap crops on Lepidopteran pests of cabbage in Hawaii. Environmental Entomology 1996;25: 39-47.

[30] Bender D A, Morrison W P and Frisbie R E. Intercropping cabbage and Indian mustard for potential control of lepidopterous and other insects. Hortscience 1999;34: 275-279.

[31] Shelton A M and Nault B A. Dead-end trap cropping: a technique to improve management of the diamondback moth, Plutella xylostella (Lepdioptera: Plutellidae). Crop Protection 2004;23: 497-503.

[32] Musser F R, Nault B A, Nyrop J P and Shelton A M. Impact of a glossy collard trap crop on diamondback moth adult movement, oviposition, and larval survival. Entomologia Experimentalis et Applicata 2005;117: 71-81.

[33] Thompson J N. Evolutionary ecology of the relationship between oviposition preference and performance of offspring in phytophagous insects. Entomologia Experimentalis et Applicata 1988;47: 3-14.

[34] Thompson J N and Pellmyr O. Evolution of oviposition behavior and host preference in Lepidoptera: Annual Review of Entomology 1991;36: 65-89.

[35] Badenes-Perez F R, Shelton A M and Nault B A. Evaluating trap crops for diamondback moth, Plutella xylostella (Lepidoptera: Plutellidae). Journal of Economic Entomology 2004;97: 1365-1372.

[36] Shinoda T, Nagao T, Nakayama M, Serizawa H, Koshioka M, Okabe H and Kawai A. Identification of a triterpenoid saponin from a crucifer, Barbarea vulgaris, as a feeding deterrent to the diamondback moth, Plutella xylostella. Journal of Chemical Ecology 2002;28: 587-599.

[37] Hoy C W. Colorado potato beetle resistance management strategies for transgenic potatoes. American Journal of Potato Research 1999;76: 215-219.

[38] Hokkanen H M T. Biological and agrotechnical control of the rape blossom beetle Meligethes aeneus (Coleoptera, Nitidulidae). Acta Entomologica Fennica 1989;53:25-29.

[39] Arker JE. Diversity by design. Exploring trap crops and companion plants to control Phyllotreta cruciferae, the crucifer flea beetle, in broccoli. Washington State University (Doctoral Dissertation); 2012.

[40] Hannunen S. Modeling the interplay between pest movement and the physical design of trap crop systems. Agricultural and Forest Entomology 2005;7: 11-20.

[41] Boucher T J, Ashley R, Durgy R, Sciabarrasi M and Calderwood W. Managing the pepper maggot (Diptera: Tephritidae) using perimeter trap cropping. Journal of Economic Entomology 2003;96: 420-432.

[42] Hoy C W, Vaughn T T and East D A. Increasing the effectiveness of spring trap crops for Leptinotarsa decemlineata. Entomologia Experimentalis et Applicata 2000;96: 193-204.

[43] Hunt DW and Whitfield AG. Potato trap crops for control of Colorado potato beetle (Coleoptera: Chrysomelidae) in tomatoes. Canadian Entomologist 1996;128: 407-412.

[44] Weber D C, Ferro D N, Buonaccorsi J, Hazzard R V. Disrupting spring colonization of Colorado potato beetle to nonrotated potato fields. Entomologia Experimentalis et Applicata 1994;73: 39-50.

[45]   Lu W, Hou M L, Wen J H and Li J W. Effects of plant volatiles on herbivorous insects.
       Plant Protection 2007;33: 7-11.

[46]   Stan F, Helen B and Rosemary H C. Companion planting – do aromatic plants dis-
       rupt host-plant finding by the cabbage root fly and the onion fly more effectively
       than non-aromatic plants? Entomologia Experimentalis et Applicata 2003;109:
       183-195.

[47]   Uvah I I I and Coaker T H. Effect of mixed cropping on some insect pests of carrots
       and onions. Entomologia Experimentalis et Applicata 1984;36: 159-167.

[48]   Isman B. Botanical insecticides, deterrents, and repellents in modern agriculture and
       increasingly regulated world. Annual Review of Entomology 2006;51: 45-66.

[49]   Anon. Companion planting. 2004a [Online]. Available: http://www.ghorganics.com/
       page2.html.

[50]   Anon. Organic vegetable IPM guide. 2004b [Online]. Available: http://msucares.com/
       pubs/publications/p2036.pdf.

[51]   Landolt P J, Hofstetter R W and Biddick L L. Plant essential oils as arrestants and re-
       pellents for neonate larvae of the codling moth (Lepidoptera: Tortricidae). Environ-
       mental Entomology 1999;28: 954-960.

[52]   Scheffler I and Dombrowski M. Behavioural responses of Blattella germanica L. (Or-
       thopt., Blattelidae) induced by repellent plant extracts. Journal of Applied Entomolo-
       gy 1993;115: 499-505.

[53]   Dabrowski Z T and Seredynska U. Characterisation of the two-spotted spider mite
       (Tetranychus urticae Koch, Acari: Tetranychidae) response to aqueous extracts from
       selected plant species. Journal of Plant Protection Research 2007;47: 113-124.

[54]   Amarawardana L, Bandara P, Kumar V, Pettersson J, Ninkovic V and Glinwood R.
       Olfactory response of Myzus persicae (Homoptera: Aphididae) to volatiles from leek
       and chives- potential for intercropping with sweet pepper. Acta Agriculturæ Scandi-
       navica B 2007;57: 87-91.

[55]   Harvey C T and Eubanks M D. Effect of habitat complexity on biological control by
       the red imported fire ant (Hymenoptera: Formicidae) in collards. Biological Control
       2004;29: 348-358.

[56]   Talekar N S, Yang J C and Lee S T. Introduction of Diadegma semiclausum to control
       diamondback moth in Tawian, p. 263-270. in Diamondback moth and other crucifer
       pests N.S. Talekar (ed.). Proc. 2nd International Workshop, Shunhua, Taiwan. Asian
       Vegetable Research and Development Center; 1992.

[57]   Little B. Companion planting in Australia. Reed Books Pty Ltd: Frenchs Forest, New
       South Wales; 1989.

[58] Philbrick H and Gregg R B. Companion plants and how to use them. The Devin-Adair Company. Old Greenwich, Connecticut; 1982.

[59] Buranday R P and Raros R S. Effects of cabbage-tomato intercropping on the incidence and oviposition of the diamond back moth, Plutella xylostella (L.) Phillippine Entomology 1975;2: 369-375.

[60] Tahvanainen J O and Root R B. The influence of vegetational diversity on population ecology of a specialize herbivore, Phyllotreta cruciferae (Coleoptera: Chrysomelidae) Oecologia 1972;10: 321-346.

[61] Dover J W. The effects of labiate herbs and white clover on Plutella xylostella oviposition. Entomologia Experimentalis Applicata 1986;42: 243-247.

[62] Latheef M A and Irwin R D. The effect of companionate planting on lepidopteran pests of cabbage. The Canadian Entomologist 1979;111: 863-864.

[63] Billiald H, Finch S, Collier R H and Elliott M. Disrupting host-plant finding by pest insects of cruciferous plants using aromatic and non-aromatic plants. IOBC/WPRS Bulletin 2005;28 (4): 53-59.

[64] Couty A, van Emden H, Perry J N, Hardie J, Pickett J A and Wadhams L J. The role of olfaction and vision in host-plant finding by the diamondback moth, Plutella xylostella. Physiological Entomology 2006;31: 134-145.

[65] Zohren E. Laboruntersuchungen zu Massenanzucht, Lebensweise, Eiablage and Eiablageverhalten der Kohlfliege, Chortophila brassicae (Bouché) (Diptera, Anthomyiidae). Zeitschrift für angewandte Entomologie 1968;62: 139-188.

[66] Finch S, Billiald H and Collier R H. Companion planting-do aromatic plants disrupt host-plant finding by the cabbage root fly and the onion fly more effectively than non-aromatic plants? Entomologia Experimentalis et Applicata 2003;109: 183-195.

[67] Poveda K, Gomez M I and Martinez E. Diversification practices: their effect on pest regulation and production. Revista Colombiana de Entomologia 2008;34: 131-144.

[68] Perrin R M and Phillips M L. Some effects of mixed cropping on the population dynamics of insect pests. Entomologia Experimentalis et Applicata 1978; 24: 385-393.

[69] Dempster J P. Some effects of weed control on the numbers of the small cabbage white (Pieris rape L.) on Brussels sprouts. Journal of Applied Ecology 1969;6: 339-345.

[70] Smith J G. Influence of crop backgrounds on aphids and other phytophagous insects on Brussels sprouts. Annals of Applied Biology 1976;83: 1-13.

[71] Theunissen J, and den Ouden H. Effects of intercropping with Spergula arvensis on pests of Brussels sprouts. Entomologia Experimentalis et Applicata 1980;27: 260.268.

[72] Kostal V and Finch S. Influence of background on host plant selection and subsequent oviposition by the cabbage root fly (Delia radicum). Entomologia Experimentalis et Applicata 1994;70: 153-163.

[73] Theunissen J, Booij C J H, Schelling G and Noorlander J. Intercropping white cabbage with clover. IOBC/WPRS Bulletin XV/ 1992.;4: 104-114

[74] Finch S and Edmonds G H. Undersowing cabbage crops with clover-the effects on pest insects, ground beetles and crop yield. IOBC/WPRS Bulletin 1994;17(8): 159-167.

[75] Tukahirwa E M and Coaker T H. Effect of mixed cropping on some insect pests of brassicas; reduced Brevicoryne brassicae infestations and influences on epigeal predators and the disturbance of oviposition behavior in Delia brassicae. Entomologia Experimentalis et Applicata 1982;32: 129-140.

[76] Theunissen J. Effects of intercropping on pest populations in vegetable crops. IOBC/WPRS Bulletin 1994;17 (8): 153-158.

[77] Rovira A A. Plant root exudates. Botanical Review 1969;35: 35-39.

[78] Matsumoto K and Kotulai J R. Field tests on the effectiveness of Azadirachta companion planting as a shoot borer repellent to protect mahogany. Japan International Research Center for Agricultural Sciences 2002;10: 1-8.

[79] Pettersson J, Ninkovic V and Ahmed E. Volatiles from different barley cultivars affect aphid acceptance of neighbouring plants. Acta Agriculturae Scandinavica B 1999;49: 152-157.

[80] Ninkovic V, Olsson U and Pettersson J. Mixing barley cultivars affects aphid host plant acceptance in field experiments. Entomologia Experimentalis et Applicata 2002;102: 177-182.

[81] Glinwood R T, Gradin T, Karpinska B, Ahmed E, Jonsson L and Ninkovic V. Aphid acceptance of barley exposed to volatile phytochemicals differs between plants exposed in daylight and darkness. Plant Signaling Behavior 2007;2: 205-210.

[82] Atsatt P R and O'Dowd D J. Plant defense guilds. Science 1976;193: 24-29.

[83] Björkman M, Hambäck P A and Rämert B. Neighbouring monocultures enhance the effect of intercropping on the turnip root fly (Delia floralis). Entomologia Experimentalis et Applicata 2007;124: 319-326.

[84] Morley K, Finch S and Collier R H. Companion planting-behavior of the cabbage root fly on host plants and non-host plants. Entomologia Experimentalis et Applicata 2005;117: 15-25.

[85] Hopkins R J, Wright F, McKinley R G and Birch A N E. Analysis of sequential behavior patterns: the oviposition decision of the turnip root fly, Delia floralis. Entomologia Experimentalis et Applicata 1996;80: 93-96.

[86] Hopkins R J, Wright F, Birch A N E and McKinlay E R G. The decision to reject an oviposition site: sequential analysis of the post-alighting behavior of Delia floralis. Physiological Entomology 1999;24: 41-50.

[87] Feeny P. Plant apparency and chemical defense, pp. 1-10. in Biochemical interaction between plants and insects, J. Wallace and R. Mansell (eds.). Recent Advances in Phytochemistry 1976;10: 1-40.

[88] Ryan J, Ryan M F and McNaeidhe F. The effect of interrow plant cover on populations of cabbage rootfly, Delia brassicae (Wied.). Journal of Applied Ecology 1980;17: 31-40.

[89] Perrin R M. Pest management in multiple cropping systems. Agro-ecosystems 1977;3: 93-118.

[90] Altieri M A and Doll J D. Some limitations of weed biocontrol in tropical ecosystems in Colombia. In: T.E. Freeman (ed.), Proceedings IV International symposium on Biological Control of Weeds. University of Florida, Gainesville pp. 74-82; 1978.

[91] Thresh M. Cropping practices and virus spread. Annual Review of Phytopathology 1982;20: 193-218.

[92] Toba H H, Kishaba A N, Bohn G W and Hield H. Protecting muskmelon against aphid-borne viruses. Phytopathology 1977;67: 1418-1423.

[93] Finch S and Kienegger M A. behavioural study to help clarify how undersowing with clover affects host plant selection by pest insects of brassica crops. Entomologia Experimentalis et Applicata 1997;84: 165-172.

[94] Cook S M, Khan Z R and Pickett J A. The use of "push-pull" strategies in integrated pest management. Annals Review Entomology 2007;52: 375-400.

[95] Khan Z R, Ampong-Nyarko K, Chilishwa P, Hassanali A, Kimani S, Lwande W, Overholt W A, Pickett J A, Smart L E, Wadhams L J and Woodcock C M. Intercropping increases parasitism of pests. Nature 1997;388: 631-632.

[96] Khan Z R and Pickett J A. The 'push-pull' strategy for stemborer management: a case study in exploiting biodiversity and chemical ecology. Ecological Engineering for Pest Management-Advances in Habitat Manipulation for Arthropods (ed. by GM Gurr, SD Wratten and MA Altieri), CABI Publishing, Wallingford, UK; 2004; pp. 155-164.

[97] Komi A, Gounou S and Tamò M. The role of maize-legumes-cassava intercropping in the management of maize ear borers with special reference to Mussidia nigrivenella Ragonot (Lepidoptera: Pyralidae).Annales de la Société entomologique de France 2006;42: 495-502.

[98] Van den Bosch, R, and Telford, A D. Environmental modification and biological control. In Biological Control of Pests and Weeds, ed. P DeBac, pp. 459-488. New York: Reinhold; 1964.

[99] Landis D A, Wratten S D and Gurr G M. Habitat management to conserve natural enemies of arthropod pests in agriculture. Annual Review of Entomology 2000;45: 175-201.

[100] Coll M and Guershon, M. Omnivory in terrestrial arthropods: mixing plant and prey diets. Annual Review of Entomology 2002;47: 267-297.

[101] Patt J M, Wainright S C, Hamilton G C, Whittinghill D, Bosley K, Dietrick J and Lashomb J H. Assimilation of carbon and nitrogen from pollen and nectar by a predaceous larva and its effects on growth and development. Ecological Entomology 2003;28: 717-728.

[102] Wäckers F L, van Rijn P C J, Bruin J. Plant-provided food for carnivorous insects: a protective mutualism and its applications. Cambridge: Cambridge University Press; 2005.

[103] Hickman J M and Wratten S D. Use of Phacelia tanacetifolia strips to enhance biological control of aphids by hoverfly in cereal fields. Journal of Economic Entomology 1996;89: 832-840.

[104] Schneider F. Beitrag zur Kenntnis der Generationsverhaltinisse und diapause raberischer Schwebfligen. Mitt. Schweiz. Entomol. Ges 1948;21: 249-285.

[105] Jervis, MA., Lee, JC. and Heimpel, GE. 2004. Conservation biological control using arthropod predators and parasitoids – the role of behavioural and life-history studies. In: Gurr G, Wratten SD, Altieri M, editors. Ecological engineering for pest management: advances in habitat manipulation of arthropods. Melbourne: CSIRO Press. p. 69-100.

[106] Lee J C, Heimpel G E and Leibee G L. Comparing floral nectar and aphid honeydew diets on the longevity and nutrient levels of a parasitoid wasp. Entomologia Experimentalis et Applicata 2004;111: 189-199.

[107] Baggen L R, Gurr G M and Meats A. Flowers in tri-trophic systems: mechanism allowing selective exploitation by insect natural enemies for conservation biological control. Entomologia Experimentalis et Applicata 1999;91: 155-161.

[108] Gurr G M and Nicol H I. Effect of food on longevity of adults of Trichogramma carverae Oatman and Pinto and Trichogramma carverae Oatman and Pinto and Trichogramma nr brassicae Bezdenko (Hymenoptera: Trichogrammatidae). Australian Journal of Entomology 2000;39: 185-187.

[109] Bickerton M W and Hamilton G C. Effects of Intercropping with flowering plants on predation of Ostrinia nubilalis (Lepidoptera: Crambidae) eggs by generalist predators in bell peppers. Environmental Entomology 2012;41: 612-620

[110] Begum M, Gurr G M, Wratten S D, Hedberg P and Nicol H I. The effect of floral nectar on the grapevine leafroller parasitoid Trichogramma carverae. International Journal of Ecology and Environmental Sciences 2004;30: 3-12.

[111] Begum M, Gurr G M, Wratten S D, Hedberg P R and Nicol H I. Using selective food plants to maximize biological control of vineyard pests. Journal of Applied Ecology 2006;43: 547-554.

[112] Lavandero B I, Wratten S D, Didham R K and Gurr G. Increasing floral diversity for selective enhancement of biological control agents: A double-edged sward? Basic and Applied Ecology 2006;7: 236-243.

[113] Wade M R and Wratten S D. Excised or attached inflorescences? Methodological effects on parasitoid wasp longevity. Biological Control 2007;40: 347-354.

[114] Berndt L A, Wratten S D and Hassan P G. Effects of buckwheat flowers on leafroller (Lepidoptera: Tortricidae) parasitoids in a New Zealand vineyard. Agriculture and Forest Entomology 2002;4: 39-45.

[115] Carreck N L and Williams I H. Observations on two commercial flower mixtures as food sources for beneficial insects in the UK. Journal of Agricultural Science 1997;128: 397-403.

[116] White A J, Wratten S D, Berry N A and Weighmann U. Habitat manipulation to enhance biological control of Brassica pests by hover flies (Diptera: Syrphidae). Journal of Economic Entomology 1995;88: 1171-1176.

[117] Freeman Long R, Corbett A, Lamb C, Reberg-Horton C, Chandler J and Stimmann M. Beneficial insects move from flowering plants to nearby crops. California Agriculture 1998;52: 23-26

[118] Nalepa C A, Bambara S B and Burroughs A M. Pollen and nectar feeding by Chilocorus kuwanae (Silvestri) (Coleoptera: Coccinellidae). Proceedings of the Entomological Society of Washington 1992;94: 596-597.

[119] Pemberton R W, Vandenberg N J. Extrafloral nectar feeding by ladybird beetles (Coleoptera: Coccinellidae). Proceedings of the Entomological Society of Washington 1993; 95: 139-151.

[120] Jervis M A. and Kidd N A C. Host-feeding strategies in hymenopteran parasitoids. Biological Reviews 1986;61: 395-434.

[121] Langenberger M W and Davis A R. Temporal changes in floral nectar production, reabsorption, and composition associated with dichogamy in annual caraway (Carum carvi; Apiacea). American Journal of Botany 2002;1588-1598.

[122] Quarles W and Grossman J. Insectary plants, intercropping and biological control. The IPM Practitioner 2002;24 (3): 1-11.

[123] Fiedler A K and Landis D A. Plant characteristics associated with natural enemy abundance at Michigan native plants. Environmental Entomology 2007;36: 878-886.

[124] Cowgill S E, Wratten S D and Sotherton N W. The effect of weeds on the numbers of hoverfly (Diptera: Syrphidae) adults and the distribution and composition of their eggs in winter wheat. Annals of Applied Biology 1993;123: 499-514.

[125] Maredia K M, Gage S H, Landis D A and Scriber J M. Habitat use patterns by the seven-spotted lady beetle (Coleoptera: Coccinellidae) in a diverse agricultural landscape. Biological Control 1992;2: 159-165.

[126] MacLeod A. Alternative crops as floral resources for beneficial hoverflies (Diptera: Syphidae). Proceedings of the Brighton Crop Protection Conference, pp. 997-1002. Brighton: British Crop Protection Council; 1992.

[127] Lövei G L, Hodgson D J, MacLeod A and Wratten S D. Attractiveness of some novel crops for flower-visiting hover flies (Diptera: Syrphidae): comparisons from two continents. Pp. 368-370 in: Pest control and sustainable agriculture. Corey, S.; Dall, D.; Milne, W. ed. Australia, CSIRO; 1993. p514.

[128] Patt J M, Hamilton G C and Lashomb J H. Foraging success of parasitoid wasps on flowers: interplay of insect morphology, floral architecture and searching behavior. Entomologia Experimentalis et Applicata 1997;83: 21-30.

[129] Von Klinger K. Auswirkungen eingesater Randstreifen an einem Winterweizen-Feld auf die Raubarthropenfuana und den Getreideblattlausbefall. Journal of Applied Entomology 1987; 104: 47-58.

[130] Ruppert V and Klingauf F. The attractiveness of some flowering plants for beneficial insects as exemplified by Syrphinae (Diptera: Syrphidae). Mitteilungen der Deutschen Gesellschaft für Allgemeine und Angewandte Entomologie 1988;6 (1-3): 255-261.

[131] Gruys P. Hits and misses. The ecological approach to pest control in orchards. Entomolgia Experimentalis et Applicatis 1982;31: 70-87.

[132] Thomas M B, Wratten S D and Sotherton N W. Creation of island habitats in farmland to manipulate populations of beneficial arthropods: predator densities and species composition. Journal of Applied Ecology 1991;28: 906-917.

[133] Dennis P and Fry G L A. Field margins: can they enhance natural enemy population densities and general arthropod diversity on farms? Agriculture Ecosystems and the Environment 1992;40: 95-115.

[134] Wyss E. The effects of artificial weed strips on diversity and abundance of the arthropod fauna in a Swiss experimental apple orchard. Agriculture Ecosystems and Environment 1996;60: 47-59.

[135] Letourneau D K and Altieri M A. Environmental management to enhance biological control in agroecosystems. In: Bellows, T.S., Fischer, T.W. (Eds.), Handbook of Biological Control. Academic Press, San Diego; 1999. p319-354.

[136] Woodcock B A, Potts S G, Pilgrim E, Ramsay A J, Tscheulin T, Parkinson A, Smith R E N, Gundrey A L, Brown V K and Tallowin J R. The potential of grass field margin management for enhancing beetle diversity in intensive livestock farms. Journal of Applied Ecology 2007; 44: 60-69.

[137] Jonsson M, Wratten S D, Landis D A and Gurr G M. Recent advances in conservation biological control of arthropods by arthropods. Biological Control 2008;45: 172-175.

[138] Hossain Z, Gurr G M, Wratten S D and Raman A. Habitat manipulation in lucerne (Medicago sativa L.): arthropod population dynamics in harvested and 'refuge' crop strips. Journal of Applied Ecology 2002;39: 445-454.

[139] Thomas M B, Wratten S D and Sotherton N W. Creation of island habitats in farmland to manipulate populations of beneficial arthropods: predator densities and species composition. Journal of Applied Ecology 1992;29: 524-531.

[140] Perdikis D, Fantinou A and Lykouressis D. Enhancing pest control in annual crops by conservation of predatory Heteroptera. Biological Control 2011;59: 13-21.

[141] Lavandero B, Wratten S D, Shishehbor P and Worner S. Enhancing the effectiveness of the parasitoid Diadegma semiclausum (Helen): movement after use of nectar in the field. Biological Control 2005;34: 152-158.

[142] Tylianakis J M, Didham R K and Wratten S D. Improved fitness of aphid parasitoids receiving resource subsidies. Ecology 2004;85: 658-666.

[143] Barbosa P. ed. Conservation Biological Control. Sand Diego, CA: Academic. 396 pp; 1998.

[144] Gurr G M, Wratten S, and Altieri M A. Ecological engineering for pest management: advances in habitat manipulation for arthropods. CABI. Wallington. CABI, Wallingford, UK; 2004.

[145] Heimpel G E and Jervis M A. Does floral nectar improve biological control by parasitoids? Plant-provided Food and Plant-Carnivore Mutualism (eds F.L. Wackers, P.C.J. Van Rijn & J. Bruin) Cambridge University Press, Cambridge, UK; 2005. p267-304.

[146] Wäckers F L, Romesis J and van Rijn P. Nectar and pollen-feeding by insect herbivores and implications for tri-trophic interactions. Annual Review of Entomology 2007;52: 301-323.

[147] Bowie M H, Wratten S D and White A J. Agronomy and phenology of "companion plants" of potential for enhancement of biological control. New Zealand Journal of Crop Horticultural Science 1995;23: 432-427.

[148] Bigger D S and Chaney W E. Effects of Iberis umbellate (Brassicaceae) on insect pests of cabbage and on potential biological control agents. Environmental Entomology 1998;27: 161-167.

[149] Finke D L and Denno R F. Spatial refuge from intraguild predation: implications for prey suppression and trophic cascades. Oecologia 2006;149: 265-275.

[150] Hughes A R and Grabowski J H. Habitat context influences predator interference interactions and the strength of resource partitioning. Oecologia 2006;149: 256-264.

[151] Haslett J R. Interpreting patterns of resource utilization: randomness and selectivity in pollen feeding by adult hoverflies. Oecologia 1989;78: 433-432.

[152] Straub C S, Finke D L and Snyder W E. Are the conservation of natural enemy biodiversity and biological control compatible goals? Biological Control 2008;45: 225-237.

[153] Risch S H. Insect herbivore abundance in tropical monocultures and polycultures: an experimental test of two hypotheses. Ecology 1981;62: 1325-1340.

[154] Prokopy R C. Integration in orchard pest and habitat management: a review. Agriculture Ecosystems and Enviroment 1994;50: 1-10.

[155] Baggen L R and Gurr G M. The influence of food on Copidosoma koehleri (Hymenoptera: encyrtidae), and the use of flowering plants as a habitat management tool to enhance biological control of potato moth, Phthorimaea operculella (Lepidoptera: Gelechiidae). Biological Control 1998;11: 9-17.

[156] Hansen L S. Introduction of Aphidoletes aphidimyza (Rond.) (Diptera: Cecidomyiidae) from an open rearing unit for the control of aphids in glasshouses. Bulletin SROP 1983; 6: 146-150.

[157] Bennison J A and Corless S P. Biological control of aphids on cucumbers: further development of open rearing units or 'bank plants' to aid establishment of aphid natural enemies. Bulletin OILB/SROP 1993;16: 5-8

[158] Van Driesche R G, Lyon S, Sanderson J P, Bennett K C, Stanek E J, Zhang R T. Greenhouse trials of Aphidius colemani (Hymenoptera: Braconidae) banker plants for control of aphids (Hemiptera: Aphididae) in greenhouse spring floral crops. Florida Entomologist 2008;91: 583-591.

[159] Stacey D L. Banker plant production of Encarsia formosa Gahan and its use in control of glasshouse whitefly on tomatoes. Plant Pathology 1997;26: 63-66.

[160] Pickett C H, Simmons G S, Lozano E and Goolsby J A. Augmentative biological control of whiteflies using transplants. Biological Control 2004;49: 665-688.

[161] Bennison J A. Biological control of aphids on cucumbers use of open rearing systems or 'banker plants' to aid establishment of Aphidius matricariae and Aphidoletes aphidimyza. Mededlingen van de Faculteit Landbouwwetenschappen. Universiteit Gent 1992;57: 457-466.

[162] Frank S D. Biological control of arthropod pests using banker plant systems: Past progress and future directions. Biological Control 2010;52: 8-16.

[163] Jacobson R J and Croft P. Strategies for the control of Aphis gossypii Glover (Hom: Aphididae) with Aphidius colemani Viereck (Hym: Braconidae) in protected cucumbers. Biocontrol Science and Technology 1998;8: 377-387.

[164] Bribosia E, Bylemans D, Migon M and Van Impe G. In-field production of parasitoids of Dysaphis plantaginea by using the rowan aphid Dysaphis sorbi as substitute host. Biological Control 2005;50: 601-610.

[165] Tooker J F and Frank S D. Genotypically diverse cultivar mixtures for insect pest management and increased crop yields. Journal of Applied Ecology 2012;49: 974-985.

[166] Letourneau D K, Armbrecht I, Rivera B S, Lerma J M, Carmona E J, Daza M C, Escobar S, Gallindo V, Gutierrez C, Lopez S D, Mejia J L, Rangel A M A, Rangel J H, Rivera L, Saavedra C A, Torres A M and Trujillo A R. Does plant diversity benefit agroecosystems? A synthetic review. Ecological Applications 2011;21: 9-21.

[167] Lin B B. Resilience in agriculture through crop diversification: adaptive management for environmental change. BioScience 2011;61: 183-193.

[168] Beizhou S, Jie Z, Jinghui H, Hongying W, Yun K and Yuncong Y. Temporal dynamics of the arthropod community in pear orchards intercropped with aromatic plants. Pest Management Science 2011;67: 1107-1114.

[169] Bone N J, Thomson L J, Ridland P M, Cole P and Hoffmann A A. Cover crops in Victorian apple orchards: effects on production, natural enemies and pests across a season. Crop Protection 2009;28: 675-683.

[170] Bandara K A N P, Kumar V, Ninkovic V, Ahmed E, Pettersson J and Glinwood R. Can leek interfere with bean plant-bean fly interaction? Test of ecological pest management in mixed cropping. Journal of Economic Entomology 2009;102 (3): 999-1008.

[171] Evans K A and Allen-Williams L J. Distant olfactory responses of the cabbage seed weevil, Ceuthorhynchus assimilis. To oilseed rape odour in the field. Physiological Entomology 1993;18: 251-256.

[172] Judd G J R and Borden J H. Distant olfactory response of the onion fly, Delia antique, to host-plant odour in the field. Physiological Entomology 1989;14: 429-441.

[173] Wigglesworth V B. The principles of insect physiology. Metheun, London, UK; 1939.

[174] Kennedy J S. Olfactory responses to distant plants and other sources. Chemical Control of Insect Behaviour: Theory and Application (ed. by HH Shorey and JJ McKelvey, Jr), John Wiley and Sons, New York, NY, USA; 1977. p67-91.

[175] Chapman R F. The Insects-Structure and Function. Hodder and Stoughton, London, UK; 1982.

[176] Dethier V G, Barton Browne L and Smith S N. The designation of chemicals in terms of the responses they elicit from insects. Journal of Economic Entomology 1960;53: 134-136.

[177] Hambäck P A, Pettersson J and Ericson L. Are associational refuges species specific? Functional Ecology 2003;17: 87-93.

[178] Bukovinszky T, Potting R P J, Clough Y, van Lenteren J C and Vet L E M. The role of pre- and post-alighting detection mechanisms in the responses to patch size by specialist herbivores. Oikos 2005;109: 435-446.

[179] Traw M R and Dawson T E. Differential induction of trichomes by three herbivores of black mustard. Oecologia 2002;131: 526-532.

[180] Soroka J J, Holowachuk J M, Gruber M Y and Grenkow L F. Feeding by flea beetles (Coleoptera: Chrysomelidae; Phyllotreta spp.) is decreased on canola (Brassica napus) seedlings with increased trichome density. Journal of Economic Entomology 2011;104: 125-136.

[181] Gao Z Z, Wu W J and Cui Z X. The Effect of Intercrop on the Densities of Phyllotreta Striolata (F.) Chinese Agricultural Science Bulletin 2004;20:214-216.

[182] Mutiga S K, Gohole L S and Auma E O. Effects of integrating companion cropping and nitrogen application on the performance and infestation of collards by Brevicoryne brassicae. Entomologia Experimentalis et Applicata 2010;134: 234-244.

[183] Bomford M K. Do tomatoes love basil but hate Brussels sprouts? Competition and land-use efficiency of popularly recommended and discouraged crop mixtures in biointensive agriculture systems. Journal of Sustainable Agriculture 2009;33: 396-417.

Chapter 3

Plutella xylostella (Linnaeus, 1758) (Lepidoptera: Plutellidae): Tactics for Integrated Pest Management in Brassicaceae

S.A. De Bortoli, R.A. Polanczyk, A.M. Vacari,
C.P. De Bortoli and R.T. Duarte

Additional information is available at the end of the chapter

1. Introduction

The diamondback moth, *Plutella xylostella* (L.) (Lepidoptera: Plutellidae), is one of the most serious pests of cultivated Brassicaceae worldwide [1,2]. This crucifer specialist may have its origin in Europe [3], South Africa [4], or East Asia [5], but is now present worldwide wherever its host plants exist [6].

In the first instar, the larvae enter into the leaf parenchyma and feed between the upper and lower surfaces of leaves creating mines. In the second instar, the larvae leave the mines, and from the second to the third instar, they feed on the leaves, destroying the leaf tissue except for the upper epidermis, leaving transparent "windows" in the leaves. Fourth-instar larvae feed on both sides of the leaves [7]. This insect has a short life cycle, around 18 days, and its population may increase up to 60-fold from one generation to the next [8]. Studies indicate that the moths can remain in continuous flight for several days while covering distances up to 1000 km per day, but how the moths survive at such low temperatures and high altitude is not known [1]. In eastern Canada, annual populations of diamondback moths originate from adult migrants from the United States [9].

P. xylostella was the first crop insect reported to be resistant to dichloro-diphenyl-trichloro-ethane (DDT), only 3 years after the start of its use [10], and subsequently it has shown significant resistance to almost every insecticide applied in the field, including new chemical compounds [11,12]. In addition, diamondback moth has the distinction of being the first insects to develop resistance in the field to the bacterial insecticide *Bacillus thuringiensis* [13,14]. The resistance of *P. xylostella* populations to *B. thuringiensis* has been observed by [15-23] in the USA (Florida, Hawaii, and New York), Central America (Mexico, Costa Rica, Guatemala, Honduras, and Nicaragua), and Asia (Japan, China, Malaysia, and the Philippines). In Brazil, [24] it was documented this pest's resistance in environments where B. thuringiensis is commonly used as a bioinsecticide.

This has prompted increased efforts worldwide to develop IPM programs for P. xylostella, based principally on new management tactics that are not yet used in the field for this pest [8,25,26]. In this chapter, we give an overview of the association of P. xylostella with its host plants and natural enemies, and describe management strategies and practices for control of the diamondback moth.

2. Tactics for integrated pest management

2.1. Biological control

Biological control can be defined as the use of one type of organism to reduce the population density of another. Biological control has been used for approximately two millennia, and has been widely used in pest management since the end of the nineteenth century [27]. The following types of biological control can be distinguished: natural, conservative, inoculative (or classical), and augmentative. Natural biological control involves the reduction of pest organisms by their natural enemies and has been occurring since the evolution of the first terrestrial ecosystems, 500 million of years ago [28]. It takes place in all of the world's ecosystems without any human intervention, and, in economic terms, is the greatest contribution of biological control to agriculture [29]. Conservation biological control consists of human actions that protect and stimulate the performance of naturally occurring enemies [30]. In inoculative biological control, natural enemies are collected in an exploration area (usually the area of origin of the pest) and then released in new areas where the pest was accidentally introduced. In augmentative biological natural control, natural enemies are mass-reared in biofactories for release in large numbers to obtain immediate pest control [28].

2.1.1. Entomophagous agents: parasitoids and predators

Parasitoids can be defined as insects that are only parasitic in their immature stages, kill their host in the process of development, and have free-living adults that do not move their hosts to nests or hideouts [31].

All stages of the diamondback moth are attacked by numerous parasitoids and predators, with parasitoids being the more widely studied. Over 90 parasitoid species attack the diamondback moth [32]. Egg parasitoids belonging to the polyphagous genera Trichogramma and Trichogrammatoidea contribute little to natural control and require frequent mass releases. Larval parasitoids are the most predominant and effective. Many of the effective larval parasitoids belong to two major genera, Diadegma and Cotesia; a few Diadromus spp., most of which are pupal parasitoids, also exercise significant control [1]. The majority of these species come from Europe where the diamondback moth is believed to have originated [1]. In countries near Brazil, such as Argentina, *P. xylostella* larval parasitoids collected in the field include the species *Diadegma insulare* (Cresson) (Hymenoptera: Ichneumonidae), *Oomyzus sokolowskii* (Kurdjumov) (Hymenoptera: Eulophidae), and *C. plutellae* (Kurdjumov) (Hymenoptera: Braconidae) [33].

Seven species of parasitoids were observed in a *P. xylostella* population on cabbage crops in the Brasilia region of Brazil, with the two most common species being *Diadegma liontiniae* (Brethes) (Hymenoptera: Ichneumonidae) and *Apanteles piceotrichosus* (Blanchard) (Hymenoptera: Braconidae). *Cotesia plutellae* (Kurdjumov) (Hymenoptera: Braconidae) and *Actia* sp., previously more abundant, had become very minor parasitoids. Six species of hyperparasitoids emerged from *D. liontiniae* and *A. piceotrichosus*, showing a high diversity of natural enemies in this region of recent colonization by *P. xylostella* [34].

In organically farmed kale in Pernambuco, Brazil, seven natural enemies of *P. xylostella* were observed: three parasitoids, *C. plutellae* Kurdjumov (Hymenoptera: Braconidae), *Conura pseudofulvovariegata* (Becker) (Hymenoptera: Chalcididae) and *Tetrastichus howardi* (Olliff) (Hymenoptera: Eulophidae), and four predators, *Cheiracanthium inclusum* (Hentz) (Araneae: Miturgidae), *Pheidole* sp. Westwood (Hymenoptera: Formicidae), and nymphs and adults of *Podisus nigrispinus* (Dallas) (Hemiptera: Pentatomidae) [35].

Several studies have been conducted in Brazil to examine whether these entomophagous agents of the diamondback moth could be used as a biological control for this pest in crucifer crops.

Parasitoids of the genus *Trichogramma* are among the entomophagous agents that have already been studied for *P. xylostella*. The species *T. pretiosum* Riley (Hymenoptera: Trichogrammatidae), Tp8 strain, can parasitize approximately 15 *P. xylostella* eggs in the first or second generation when reared in this host under laboratory conditions, with 100% emergence, and 10 to 11 days for adult emergence [36]. Eggs of two *P. xylostella* populations, one reared on kale leaves and the other on broccoli leaves, were exposed to the *T. pretiosum* Tp8 strain, and the number of parasitized eggs was 5.8–9.4 on kale and 3.2–8.4 on broccoli [37]. Furthermore, the optimal way to mass rear this parasitoid in the laboratory is to use eggs glued to blue, green, or white colored cards [37].

The impact on non-target species, particularly *Trichogramma*, of insecticides for *P. xylostella* control should be analyzed because some are toxic to these parasitoids in crucifers. Endosulfan and etofenprox, classified as class-4 toxic products, are extremely toxic to the parasitoids. Triflumuron, classified as a non-toxic product, is selective for these parasitoids in the eggs of *P. xylostella* [26]. The combination of chemicals or natural insecticidal products from vegetables with certain cultivars of crucifers enables more effective management of the diamondback moth, particularly in the case of the interaction between pyroligneous extract and cabbage. However, the interaction among cultivars and products can be detrimental to the effectiveness of *T. pretiosum* and *T. exiguum*, and thus requires a careful evaluation to minimize the impact on these natural enemies [38]. Bioinsecticides based on *B. thuringiensis* for controlling *P. xylostella* can influence the parasitoid *T. pretiosum* in the moth's eggs. The application of isolates of *B. thuringiensis* on *P. xylostella* larvae influenced the parasitism of *T. pretiosum* in eggs of subsequent pest generations [39].

Another parasitoid of *P. xylostella* larvae, which has been studied in Brazil, is *O. sokolowskii*. The duration of the immature stage of these parasitoids can range from 12.9 to 31.6 days at 28 and 18°C, respectively, and the number of adults emerged per pupa of *P. xylostella* varies between 7.3 and 12, with a sex ratio of between 0.86 and 0.91 [40]. During a year, the number of generations of *O. sokolowskii* is always higher than that of *P. xylostella*, suggesting that *O. sokolowskii* could develop up to 24 generations per year while the diamondback moth could reach 20 annual generations [40]. Furthermore, the *O. sokolowskii* parasitoid is able to disperse and parasitize *P. xylostella* throughout a kale field up to 24 meters from the release point [41].

Another larval parasitoid studied in Brazil for *P. xylostella* is *A. piceotrichosus*, which was collected in the Rio Grande do Sul State. Its immature stage was observed to last 14.6 to 15.5 days and its adult longevity was found to be 12.7 to 13.4 days [42].

Among the stink bug predators, *P. nigrispinus* has great potential for use in *P. xylostella* control. *P. nigrispinus* has been reported preying on *P. xylostella* in crucifer crops [35], and, furthermore, this predator consumed on average 10.9 larvae or 5.5 pupae in 24 h [43]. Adults of *Orius insidiosus* (Say) (Hemiptera: Anthocoridae) has been reported consuming 5.9 diamondback moth eggs in 24 h [44].

2.1.2. Entomopathogens: Bacteria

The occurrence of *P. xylostella* populations of resistant to certain active ingredients, like synthetic and biological insecticides, has caused a considerable increase in research directed at developing tactics for Integrated Pest Control based on economic, social, and ecological parameters [21,45-47).

Recent studies on control strategies and population reduction of *P. xylostella* using microorganisms has been increasingly cited in the scientific community, with emphasis on the entomopathogenic bacterium *B. thuringiensis* Berliner (1911) [48-51,39].

This entomopathogen can be easily found in different environments [52,53], and it is characterized by a variety of strains, each forming one or more protein crystals (Cry) and cytolytic toxins [54] that have insecticidal activity and determined its efficiency as a control on certain agricultural pests. Another type of insecticidal protein that can be synthesized by some strains of *B. thuringiensis* is "Vegetative Insecticidal Proteins" (Vip), whose insecticide action spectrum operates in different insect species [55].

A long history of intensive research has established that their toxic effect is due primarily to their ability to form pores in the plasma membrane of the midgut epithelial cells of susceptible insects [56,57]. The presently available information still supports the notion that *B. thuringiensis* Cry toxins act by forming pores, but most events leading to their formation, following binding of the activated toxins to their receptors, remain relatively poorly understood [58].

Strains of *B. thuringiensis* can produce from one to five toxins that represent a large variability in toxicity and interfere in the expression levels and the spectrum control of insects, and differ in their specificity to certain species [59]. For example, the Cry proteins are show high toxicity to insects of the orders Lepidoptera, Coleoptera, Hymenoptera, Diptera, Orthoptera, and Mallophaga, and to other organisms such as nematodes and mites [60,54,61].

Among the different protein crystals identified in insect control, 59 toxins were tested against 71 Lepidoptera species [62]. The broadest range of toxins was tested against *P. xylostella* (43 toxin types), which was one of only 12 species that were tested against 15 toxins or more [62].

In Brazil, *P. xylostella* is controlled using entomopathogenic bacteria in phytosanitary applications of formulation products properly registered for a particular crop, most commonly biological products containing *B. thuringiensis* var. *kurstaki*, which expresses Cry1Aa, Cry1Ab, and Cry1Ac toxins [49] (Table 1).

Crops	Commercial Products	Chemical Groups	Active Ingredients	Formulation	Class Toxicological	Class Environmental
Broccoli	Bac Control	Biological	*B. thuringiensis* subsp. *kurstaki*	WP	IV	IV
	Dipel	Biological	*B. thuringiensis* subsp. *kurstaki*	WP	II	IV
	Thuricide	Biological	*B. thuringiensis* subsp. *kurstaki*	WP	IV	IV
Cauliflower	Bac Control	Biological	*B. thuringiensis* subsp. *kurstaki*	WP	IV	IV
	Dipel	Biological	*B. thuringiensis* subsp. *kurstaki*	WP	II	IV
	Thuricide	Biological	*B. thuringiensis* subsp. *kurstaki*	WP	IV	IV
Cabbage	Able	Biological	*B. thuringiensis* subsp. *kurstaki*	SC	III	IV
	Agree	Biological	*B. thuringiensis* subsp. *aizawai* GC 91 + *B. thuringiensis* subsp. *kurstaki*	WP	III	IV
	Bac Control	Biological	*B. thuringiensis* subsp. *kurstaki*	WP	IV	IV
	Dipel	Biological	*B. thuringiensis* subsp. *kurstaki*	WG	II	IV
	Dipel	Biological	*B. thuringiensis* subsp. *kurstaki*	WP	II	IV
	Thuricide	Biological	*B. thuringiensis* subsp. *kurstaki*	WP	IV	IV
	Xentari	Biological	*B. thuringiensis* subsp. *aizawai*	WG	II	III
Kale	Bac Control	Biological	*B. thuringiensis* subsp. *kurstaki*	WP	IV	IV
	Dipel	Biological	*B. thuringiensis* subsp. *kurstaki*	WP	II	IV
	Thuricide	Biological	*B. thuringiensis* subsp. *kurstaki*	WP	IV	IV

Source: [63]. WP = Wettable Powder; WG = Water-Dispersible Granules; SC = Suspension concentrate.

Table 1. Commercial products based on *Bacillus thuringiensis* recommended for controlling the population of *Plutella xylostella* in different brassica crops.

However, the low variability in the number of toxins related to formulated biological products, combined with a high number of applications in the field, puts selection pressure on the population of *P. xylostella* and, consequently, expression of resistance of this pest to protein crystals has been observed since the 1990s [20,24].

The development of resistance in *P. xylostella* populations is related to the binding of these toxins with the intestinal epithelium, which occurs through the same membrane receptors [19,22].

Some alternative methods of resistance management of this pest towards *B. thuringiensis* toxins can reduce resistance and even make it possible to break the resistance to biological products [22,64].

According to [49], mixed formulations of different bacteria or isolates of *B. thuringiensis* that have a wide variety of *Cry* toxins, organized in isolation or together, have the ability to reduce selection pressure and, consequently, the development of new cases of resistance in populations of *P. xylostella*.

To improve the biological control of *P. xylostella* using this entomopathogenic bacterium, several studies have initially focused to on the characterization of new strains of *B. thuringiensis*, with the objective of discovering more efficient insecticides and implementing them in new formulations [65,66,51]. In a study conducted by [49] using stored grains and different strains of *B. thuringiensis* from soils of several regions of Brazil, there was high mortality (98–100%) of second-instar larvae of *P. xylostella*. These results have demonstrated that a high variability of *Cry* genes in the same strain can constitute a substantial tool for resistance management of this pest, with subsequent use in the synthesis of new biological products.

In pathogenicity tests, the strains behave in different ways, and few of them are able to cause total mortality in the insects analyzed. In research conducted by [51], approximately 19% of the strains tested caused total mortality to second-instar larvae of *P. xylostella* between 24 and 48 hours.

In this case, in addition to pathogenicity and virulence tests, researchers should analyze the sublethal effects of these strains on the remaining individuals, an important parameter in the toxicological evaluation of *B. thuringiensis* strains [67,68].

Many biological characteristics of *P. xylostella* may be influenced by the sublethal effects of these toxins, causing discernible changes in insect behavior, such as appetite loss, decreased movement with subsequent paralysis, change in the tegument color from bright green to dark yellow or dark brown, and loss of reaction to touch [69,51].

According to [51] and [8], the most pronounced biological changes observed between phytosanitary applications with strains and commercial products based on *B. thuringiensis* were in the viability of larvae and pupae and the weight of pupae. The biological characteristics less influenced by these strains were related to the caterpillar and pupal period and sex ratio [8].

The behavior of strains or commercial products based on *B. thuringiensis* that result in individuals surviving phytosanitary application, but that provide sublethal effects in subsequent generations, may be a significant tool for Integrated Pest Management [8], the objective of which is to improve management of the pest through interactions with other control methods, such as biological control with predators and parasitoids, which will reduce the population density due to sublethal effects caused by strains of *B. thuringiensis*. The remaining pests may be a food source and host for other insects considered beneficial to agriculture, and can help maintain and assist the populations of these arthropods in different crops.

The Integrated Management of *P. xylostella* based on biological control with the entomopathogenic bacterium *B. thuringiensis* is an important method for reducing the population density of this pest in brassica crops. However, the use of this control must be well planned, because there are populations of this pest resistant to biological products, necessitating the use of certain methods of resistance management to eliminate these harmful individuals and, perhaps, prevent future problems with the development of resistant populations that can undermine the whole program of rational control of this pest.

2.1.3. Entomopathogens: Fungi

There is no fungus-based bioinsecticide registered for crucifer crops in Brazil; however, some entomopathogenic fungi have been studied to determine their potential as a biological control agent for P. xylostella. Among the fungi that have been studied for their activity against P. xylostella, Paecilomyces tenuipes caused the highest mortality to third-instar P. xylostella larvae, with an LC_{50} of 1.09×10^6 spores/mL at 25°C [70].

The most active crude protein extract, isolated from the CNZH strain of Isaria fumosorosea, produced 83.3% mortality in third-instar larvae 6 days post treatment [71]. Furthermore, it has been found that a synergism exists between the fungus I. fumosorosea and a plant secondary chemical, and that larval deaths were directly related to the concentration of each component in the mixtures and their cumulative effect was evident for an extended period [72].

In addition to the species already mentioned, the fungi Metarhizium anisopliae and Beauveria bassiana were also studied for the control of P. xylostella. Several strains from Benin were tested and found to cause 94% larval mortality [73]. Strains obtained in Brazil have also been tested and caused mortality to P. xylostella larvae ranging from 70% to 96% [74].

2.1.4. Entomopathogens: Nematodes

Research on the control of Lepidoptera with entomopathogenic nematodes has focused on the diamondback moth [75]. Field studies on cabbage in Java (Indonesia) confirmed that Steinernema carpocapsae can be used as a substitute for ineffective chemical insecticides [76]. Diamondback moth eggs are deposited and the emerging larvae feed on the underside of the leaves. The control of young caterpillars with entomopathogenic nematodes can therefore be optimized by directing the nematode spray to the lower side of the leaves [75]. The use of a surfactant for lowering the surface tension and of a polymer for increasing the viscosity significantly improved nematode performance against P. xylostella [77]. The performance of these adjuvants is, however, influenced by the spray application technique [75].

2.2. Chemical control

The chemical control method, recommended as one of the tools or tactics of Integrated Pest Management, is still the main strategy for reducing pest populations among crucifer producers. This preference is due to the practicality, speed, and efficiency of controlling insects considered agricultural pests, particularly P. xylostella [78].

The chemical groups used to control this pest have great variability in terms of the active ingredient, formulation, and toxicological and environmental classes (Table 2).

Crops	Chemical Group	Active Ingredient	Formulation	Class	
				Toxicological	Environmental
Broccoli	Pyrethroid	Deltamethrin	EC	III	I
	Oxime Methylcarbamate	Methomyl	SL	I	II
	Organophosphate	Acephate	SP	II	III
Canola	Pyrethroid	Bifenthrin	EC	II	II
Cauliflower	Pyrethroid	Deltamethrin	EC	III	I
	Pyrethroid	Permethrin	EC	III	II
	Organophosphate	Acephate	SP	II	III
	Naphthyl Methylcarbamate	Carbaryl	WP	III	II
Cabbage	Anthranilamide+Pyrethroid	Clorantraniliprole+Lambda-cyhalothrin	SC	II	I
	Tetranortriterpenoid	Azadirachtin	EC	III	IV
	Benzoylurea	Teflubenzuron	SC	IV	II
	Benzoylurea	Lufenuron	EC	IV	II
	Benzoylurea	Novaluron	EC	III	II
	Pyrethroid	Deltamethrin	EC	III	I
	Pyrethroid	Permethrin	EC	I	II
	Benzofuranil Methylcarbamate	Carbofuran	GR	III	II
	Oxime Methylcarbamate	Methomyl	SL	I	II
	Organophosphate	Acephate	SP	II	III
	Analog of Pyrazol	Chlorfenapyr	SC	III	II
	Phenilthiourea	Diafenthiuron	WP	I	II
	Anthranilamide	Chlorantraniliprole	SC	III	II
	Oxadiazine	Indoxacarb	WG	I	III
	Naphthyl Methylcarbamate	Carbaryl	WP	III	II
	Spinosyns	Spinosad	SC	IV	III
Kale	Pyrethroid	Deltamethrin	EC	III	I
	Oxime Methylcarbamate	Methomyl	SL	I	II
	Organophosphate	Acephate	SP	II	III
	Pyrethroid	Permethrin	EC	III	II

Source: [63]. EC = Emulsion Concentrate; SL = Soluble Concentrate; SP = Soluble Powder; WP = Wettable Powder; SC = Suspension concentrate; GR = Granules; WG = Water-Dispersible Granules.

Table 2. Chemical groups and active ingredients registered for *Plutella xylostella* control in different brassica crops.

Among the pesticides recommended for different brassicas, the chemical group of pyrethroids represents one of the most important for *P. xylostella* control. Chemical control of *P. xylostella* using a synthetic pyrethroid is recommended when larval density exceeds an economic threshold, which varies in relation to the growth stage of the crop and environmental conditions [79,80]. However, the inappropriate use of these chemical products has considerably increased the frequency of resistance in different diamondback moth populations to some types of active ingredients of this chemical group [81,82,24,83]. According to [84] and [82], *P. xylostella* populations are considered very prone to developing resistance to some active ingredients. In addition to lowering the pesticide efficiency, increasing the frequency of application may not lead to a significant reduction in crop damage.

This may be due to the biological characteristics of this species, the life cycle of which is short when compared to that of other insects, and to the cultural practice of constantly applying pesticides with the same active ingredients in more concentrated doses, without providing a chemical molecule rotation or an appropriate dosage as listed on the label of the phytosanitary product used [24].

In the context of Integrated Pest Management, cultural, physical, plant resistance, biological, and chemical control methods may be important strategies in the success of the P. *xylostella* control program [85]. Techniques such as crop residue removal, management of the interval between crops, use of tolerant cultivars, use of sprinkler irrigation, application of plant and biological products and reduction in the number of pesticide applications by measuring the economic injury level, used harmoniously and consciously, can provide significant improvement in the quality of products and the system in which the culture is embedded [86-90,83].

After a rational application of chemical controls, the first response observed in the field is the high larval mortality of P. *xylostella* in direct proportion to the commercial product concentration recommended for the determined culture [91,83]. Another response to phytosanitary application is a significant alteration in the life cycle of the insect, principally the larval period, because many chemical compounds present in insecticides affect the process of ecdysis, interfering with the transition between instars, and thereby act as a growth regulator [83].

Among the types of insecticides recommended for the control of P. *xylostella*, growth regulators have been found to have low interference with the activity of predators, parasitoids, and entomopathogenic fungi, because they do not affect the embryogenesis and reproduction of this pest, which is important since the parasitoid larvae live inside the pest's eggs before emerging as adults [85,90,38]. This is important principally because the physiological selectivity of this chemical group makes them more toxic to the pest than to the biological control agent [92,93,38,94,26].

Insecticides of plant origin are also a very important group for the population management of this pest. Among these, neem extract (*Azadirarachta indica*) has shown significant results in the control of P. *xylostella*, affecting the growth, larval mortality during ecdysis, oviposition, deformation in pupae and adults, and the physiological processes of reproduction, such as inadequate egg maturation and infertility, that interfere with larval hatching [95,90,83,38,96].

In this context, managing the population of P. *xylostella* using chemical control methods can be a very interesting strategy if well used, because of the large number of chemical groups with different active ingredients, which enables a chemical molecule rotation and prevents the development of resistance. These products can be used with other control techniques to reduce the number of applications of pesticide and improve the quality of the final product. Another very important consideration in choosing the chemical product is its selectivity, because many chemicals have high selectivity for the host but not for biological control agents, which contributes to the maintenance of populations considered beneficial to the integrated management of P. *xylostella*.

2.3. Plant resistance

The crop forms a template for various interactions between pests and their environment, and varietal resistance to pests is a key component for stabilizing an IPM system [97].

Plants have a bewildering array of responses to herbivory, broadly categorized as direct and indirect defenses and tolerance [98]. Some primary wax components, including specific long-chain alkyl components, have allelochemical activity that influences the host accept-ance behavior of *P. xylostella* larvae [99]. Furthermore, glucosinolates, a category of secon-dary products, are found primarily in species of the Brassicaceae. When tissue is damaged, for example by herbivory, glucosinolates are degraded in a reaction catalyzed by thiogluco-sidases, called myrosinases, which are also present in these species. This causes the release of toxic compounds such as nitriles, isothiocyanates, epithionitriles and thiocyanates. The glucosinolate-myrosinase system is generally believed to be part of the plant's defense against insects, and possibly also against pathogens [100].

Among various cultivars of crucifers observed, the cabbage 'chato de quintal' showed a high level of the substance glucobrassicin, and was classified as moderately resistant to *P. xylostella* [101].

Several studies have been conducted in Brazil to determine the crucifer cultivars resistant to *P. xylostella* for use in the management of this pest. Among the crucifers that are marketed in Brazil—cabbage cultivars, broccoli, kale, and cauliflower—cabbage cultivars were more re-sistant, and kale cultivars were more susceptible to diamondback moth [8]. When compared only cultivars of kale, it was found that 'Ribeirão Pires I-2620' was the most susceptible to two generations of diamondback moth [102].

The use of silicon in the integrated management of diamondback moths may help to reduce the use of pesticides. Silicon damages the jaws of larvae, limiting ingestion and causing high mortality [103].

2.4. Cultural control

The current pest management tactics pursued by growers focus on the protection of crucifer seedlings, using both cultural and chemical means, in some seasons in the established crops [104]. Because of the failure of insecticides to control the diamondback moth, interest is growing in the use of cultural controls in commercial crucifer production. Some of the classi-cal control measures that have been tried with some success are intercropping, use of sprin-kler irrigation, trap cropping, crop cover rotation, and clean cultivation [1].

The mortality of *P. xylostella* was significantly higher with the intercropping of Chinese cab-bage (*Brassica chinensis*) with garlic (*Allium sativum*) and lettuce (*Lactuca sativa*) than in mon-ocultures of Chinese cabbage. These results suggest that intercropping can suppress the diamondback moth populations for a long period rather than just the short term [105]. Fur-thermore, studies conducted in Brazil of the intercropping of cabbages with other crop plants (cabbage and green onion, cabbage and cilantro, and cabbage, green onion, and cilan-tro) did not reduce the rate of parasitism of *P. xylostella* larvae by *O. sokolowskii*, which makes it promising for diamondback moth biological control; however it did not interfere with cabbage colonization by the diamondback moth [41].

A study investigating the impact of irrigation systems on diamondback moth infestation in cabbage noted that when irrigation water was applied by sprinkler-irrigation, diamondback moth infestations were reduced by 37.5–63.9% compared with a drip-irrigated control [106].

Glucosinolates are biologically active natural products characteristic of crucifers, and cruci-fer-specialist insect herbivores, such as *P. xylostella*, frequently use glucosinolates as oviposi-tion stimuli. Benzylglucosinolate-producing tobacco plants were more attractive for oviposition by female *P. xylostella* than wild-type tobacco plants. As newly hatched *P. xylos-tella* larvae were unable to survive on tobacco, these results represent a proof-of-concept strategy for rendering non-host plants attractive for oviposition by specialist herbivores with the long-term goal of generating efficient dead-end trap crops for agriculturally impor-tant pests [107].

With regard to crop cover for crucifers, a broccoli cover-cropping system (cereal rye) result-ed in fewer leaves, smaller plants, and a slightly reduced yield when compared to the other systems. Strip-cropping broccoli with potatoes did not convey any agronomic advantages. Gross margin analysis revealed that on a total system basis, a 2.2% yield improvement or a 7% price premium was required to make the cover crop system perform as well as conven-tional practice [108].

Another study looked at the effect of two diversification strategies, one a broccoli/potato (*Solanum tuberosum*) strip crop comprising 1.65-m (tractor width) replications of two rows of potatoes and two rows of broccoli, and the other a cereal rye (*Secale cereale*) cover crop, which formed a sacrificial planting that was killed and rolled flat to minimize weed compe-tition and improve the agronomic performance of the subsequent broccoli crop. In this case, it was observed that *P. xylostella* eggs, and the subsequent larvae and pupae, were less abun-dant on broccoli with the cover crop, probably due to interference with host location and oviposition processes. The strip crop had no effect on broccoli crop yield [109].

2.5. Sex pheromones

The potential for using synthetic sex pheromone traps as a simple and practical method of monitoring population densities of insect pests has been investigated in many crop systems. Sex pheromones of *P. xylostella* have already been synthesized for use in the management of this pest in crucifers [110]. Thus, trap catches can be used to forecast infestations during pe-riods that coincide with high *P. xylostella* infestations [111].

Currently, pheromone-baited traps in the Prairie Pest Monitoring Network are used to de-tect and survey [112] the arrival of migrating moths. Recent research has shown that capture of male moths in pheromone-baited traps in the Prairie Pest Monitoring Network is correlat-ed with moderate, but not low, densities of the immature stages of the diamondback moth sampled in the same fields [113]. Then exists the potential to develop commercially available pheromone-baited traps as tools that can predict the ephemeral nature of diamondback moth population densities in the prairies and inform producers of key thresholds and tim-ing for control efforts [113].

When placed on Delta sticky traps, the artificial sex pheromone Bioplutella, marketed in Brazil, efficiently captured males of the diamondback moth and could be used for monitor-ing this pest [114].

3. Final remarks

As shown above, the management of pests on crucifers in Brazil has largely been dependent on synthetic pesticides, used prophylactically or in response to *P. xylostella* occurrence, although cultural practices have also played some role in the control of the diamondback moth [104]. The general lack of understanding of interactions between crucifers and their invertebrate pests and between pests and their natural enemies has resulted in a lack of alternative integrated options for growers. Growers would rely less heavily on the prophylactic and reactive application of broad-spectrum pesticides if they were provided with knowledge and training in identifying natural enemies and using economic thresholds. Furthermore, we again emphasize glucosinolates, their breakdown products, and plant volatile compounds as key components in these processes [115], which have been considered beneficial in the past and hold great promise for the future of integrated pest management.

Author details

S.A. De Bortoli, R.A. Polanczyk, A.M. Vacari, C.P. De Bortoli and R.T. Duarte

Department of Plant Protection, FCAV-UNESP, Jaboticabal, Sao Paulo, Brazil

References

[1] Talekar NS, Shelton AM. Biology, ecology, and management of the diamondback moth. Annual Review of Entomology 1993;38(1): 275–301.

[2] Sarfraz M, Dosdall LM, Keddie BA. Diamondback moth-host plant interactions: implications for pest management. Crop Protection 2006;25(7): 625–639.

[3] Hardy JE. *Plutella maculipennis* Curt., its natural and biological control in England. Bulletin of Entomological Research 1938;29(4): 343–372.

[4] Kfir, R. Origin of the diamondback moth (Lepidoptera: Plutellidae). Annals of the Entomological Society of America 1998;91(2): 164–167.

[5] Liu S, Wang X, Guo S, He J, Shi Z. Seasonal abundance of the parasitoid complex associated with the diamondback moth, *Plutella xylostella* (Lepidoptera: Plutellidae) in Hangzhou, China. Bulletin of Entomological Research 2000;90(3): 221–231.

[6] Shelton AM. Management of the diamondback moth: déjà vu all over again? In: Endersby NM, Ridland PM. (eds.) Proceedings of the fourth international workshop on the management on the diamondback moth and other crucifer pests. Melbourne, Victoria, Australia; 2001. p3–8.

[7] Castelo Branco M, França FH, Villas Boas GL. Traça-das-crucíferas (*Plutella xylostella*). Brasília: Embrapa Hortaliças; 1997. 4p.

[8] De Bortoli SA, Vacari AM, Goulart RM, Santos RF, Volpe HXL, Ferraudo AS. Capacidade reprodutiva e preferência da traça-das-crucíferas para diferentes brassicáceas. Horticultura Brasileira 2011;29(2): 187–192.

[9] Smith DB, Sears MK. Evidence for dispersal of diamondback moth, *Plutella xylostella* (L.) (Lepidoptera: Plutellidae), into southern Ontario. Proceedings of the Entomological society of Ontario, Canada. 1982;113: 21–27.

[10] Ankersmit GW. DDT resistance in *Plutella maculipennis* (Curt.) Lepidoptera in Java. Bulletin of Entomological Research 1953;44: 421–425.

[11] Sarfraz M, Keddie BA. Conserving the efficacy of insecticides against *Plutella xylostella* (L.) (Lepidoptera: Plutellidae). Journal of Applied Entomology 2005;129(3): 149–157.

[12] Ridland PM, Endersby NM. Some Australian populations of diamondback moth, *Plutella xylostella* (L.) show reduced susceptibility to fipronil. In: Srinivasan R, Shelton AM, Collins HL. (eds.) Sixth international workshop on management of the diamondback moth and other crucifer insect pests. Nakhon Pathom, Thailand; 2011. p21–25.

[13] Kirsch K, Schmutlerer H. Low efficacy of a *Bacillus thuringiensis* (Berl.) formulation in controlling the diamondback moth *Plutella xylostella* (L.), in the Philippines. Journal of Applied Entomology 1988;105(1-5): 249–255.

[14] Tabashnik BE, Cushing NL, Finson N, Johnson MW. Field development of resistance to *Bacillus thuringiensis* in diamondback moth (Lepidoptera: Plutellidae). Journal of Economic Entomology 1990;83(5): 1671–1676.

[15] Gong Y, Wang C, Yang Y, Wu S, Wu Y. Characterization of resistance to *Bacillus thuringiensis* toxin Cry1Ac in *Plutella xylostella* from China. Journal of Invertebrate Pathology 2010;104(2): 90–96.

[16] Sayyed AH, Gatsi R, Palacios SI, Escriche B, Wright DJ, Crickmore N. Common, but complex, mode of resistance of *Plutella xylostella* to *Bacillus thuringiensis* toxins Cry1Ab and Cry1Ac. Applied of Environmental Microbiology 2005;71(11): 6863–6869.

[17] Zhao JZ, Collins HL, Tang JD, Cao J, Earle ED, Roush RT, Herrero S, Escriche B, Ferré J, Shelton AM. Development and characterization of diamond back moth resistance to transgenic broccoli expressing high levels to Cry1C. Applied and Environmental Microbiology 2000;66(9): 3784–3789.

[18] Díaz-Gomez O, Rodríguez JC, Shelton AM, Lagunes A, Bujanos R. Susceptibility of *Plutella xylostella* (L.) (Lepidoptera: Plutellidae) populations in Mexico to commercial formulations of *Bacillus thuringiensis*. Journal of Economic Entomology 2000;93(3): 963–970.

[19] Tang JD, Shelton AM, Van Rie J, De Roeck S, Moar WJ, Roush RT, Peferoen M. Toxic-
      ity of *Bacillus thuringiensis* spore and crystal protein to resistant diamondback moth
      (*Plutella xylostella*). Applied and Environmental Microbiology 1996;62(2): 564–569.

[20] Tabashnik BE. Evolution of resistance to *Bacillus thuringiensis*. Annual Review of En-
      tomology 1994;39(1) 47–79.

[21] Perez CJ, Shelton AM. Resistence of *Plutella xylostella* (Lepidoptera: Plutellidae) to *Ba-
      cillus thuringiensis* Berliner in Central America. Journal of Economic Entomology
      1997;90(1): 87–93.

[22] Wright DJ, Iqbal M, Granero F, Ferre J. A change in a single midgut receptor in the
      diamondback moth (*Plutella xylostella*) is only part responsible for field resistance to
      Bacillus thuringiensis subsp. *kurstaki* and *Bacillus thuringiensis* subsp. *aizawai*. Applied
      and Environmental Microbiology 1997;63(5): 1814–1819.

[23] Ferré J, Real MD, van Rie J, Jansens S, Peferoen M. Resistance to the *Bacillus thurin-
      giensis* bioinsecticide in the field population of *Plutella xylostella* is due to a change in
      a midgut membrane receptor. Proceeding of the National Academic of Science
      1991;88(12): 5119–5123.

[24] Castelo Branco M, França FH, Pontes LA, Amaral PST. Avaliação da suscetibilidade a
      inseticidas em populações da traça-das-crucíferas de algumas áreas do Brasil. Horti-
      cultura Brasileira 2003;21(3): 549–552.

[25] Sarfraz M, Dosdall LM, Keddie BA. Influence of the herbivore host's wild food plants
      on parasitism, survival and development of the parasitoid *Diadegma insulare*. Biologi-
      cal Control 2012;62(1): 38–44.

[26] Goulart RM, Volpe HXL, Vacari AM, Thuler RT, De Bortoli SA. Insecticide selectivity
      to two species of *Trichogramma* in three different hosts, as determined by IOBC/
      WPRS methodology. Pest Management Science 2012;68(2): 240–244.

[27] van Lenteren J, Godfray HCJ. Europen in science in the Enlightenment and the dis-
      covery of the insect parasitoid life cycle in The Netherlands and Great Britain. Bio-
      logical Control 2005;32(1): 12–24.

[28] van Lenteren, J. The state of commercial augmentative biological control: plenty of
      natural enemies, but a frustrating lack of uptake. BioControl 2012;57(1): 1–20.

[29] Waage JK, Greathead DJ. Biological Control: challenges and opportunities. Philo-
      sophical Transactions of the Royal Society of London 1988;318(1189): 111–128.

[30] Gurr GM, Wratten SD. Measures of success in biological control. Dordrecht: Kluwer
      Academic Publishers; 2000, p430.

[31] Godfray HCJ. Parasitoids: behavioural and evolutionary ecology. New Jersey:
      Princeton University Press; 1994, p488.

[32] Goodwin S. Changes in the numbers in the parasitoid complex associated with the diamondback moth, *Plutella xylostella* (L.) (Lepidoptera) in Victoria. Australian Journal of Zoology 1979;27(6): 981–989.

[33] Bertolaccini I, Sánchez DE, Arregui MC, Favoro JC, Theiler N. Mortality of *Plutella xylostella* (Lepidoptera, Plutellidae) by parasitoids in the Province of Santa Fe, Argentina. Revista Brasileira de Entomologia 2011;55(3): 454–456.

[34] Guilloux T, Monnerat R, Castelo-Branco M, Kirk A, Bordat D. Population dynamics of *Plutella xylostella* (Lep., Yponomeutidae) and its parasitoids in the region of Brasilia. Journal of Applied Entomology 2003;127(5): 288–292.

[35] Silva-Torres CSA, Pontes IVAF, Torres JB, Barros R. New records of natural enemies of *Plutella xylostella* (L.) (Lepidoptera: Plutellidae) in Pernambuco, Brazil. Neotropical Entomology 2010;39(5): 835–838.

[36] Volpe HXL, De Bortoli AS, Thuler RT, Viana CLTP, Goulart RM. Avaliação de características biológicas de *Trichogramma pretiosum* Riley (Hymenoptera: Trichogrammatidae) criado em três hospedeiros. Arquivos do Instituto Biológico 2006;73(3): 311–315.

[37] Magalhães GO, Goulart RM, Vacari AM, De Bortoli SA Parasitismo de *Trichogramma pretiosum* Riley, 1879 (Hymenoptera: Trichogrammatidae) em diferentes hospedeiros e cores de cartelas. Arquivos do Instituto Biológico 2012;79(1): 55–90.

[38] Thuler RT, De Bortoli SA, Goulart RM, Viana CLTP, Pratissoli D. Interação tritrófica e influência de produtos químicos e vegetais no complexo: brássicas x traça-das-crucíferas x parasitóides de ovos. Ciência e Agrotecnologia 2008;32(4): 1154–1160.

[39] De Bortoli AS, Vacari AM, Magalhães GO, Dibelli W, De Bortoli CP, Alves MP. Subdosagens de *Bacillus thuringiensis* em *Plutella xylostella* (Lepidoptera: Plutellidae) e *Trichogramma pretiosum* (Hymenoptera: Trichogrammatidae). Revista Caatinga 2012;25(2) 50–57.

[40] Ferreira SWJ, Barros R, Torres JB. Exigências térmicas e estimava do número de gerações de *Oomysus sokolowshii* (Kurdjumov) (Hymenoptera: Eulophidae), para regiões produtoras de crucíferas em Pernambuco. Neotropical Entomology 2003;32(3): 407–411.

[41] Silva-Torres CSA, Torres JB, Barros R. Can cruciferous grown under variable conditions influence biological control of *Plutella xylostella* (Lepidoptera: Plutellidae)? Biocontrol Science and Technology 2011;21(6): 625–641.

[42] Gonçalves RR, Di Mare RA. Biologia da traça das crucíferas, *Plutella xylostella* Linnaeus (Lepidoptera: Yponomeutidae), sob condições controladas de temperatura, e parasitoides assiciados. Parte III. Estudo sobre a biologia de *Apanteles piceotrichosus* (Blanchard) (Hymenoptera: Braconidae): análise do efeito de endocruzamento. Revista Brasileira de Zoologia 2005;22(3): 806–809.

[43] Vacari AM, De Bortoli SA, Torres JB. Relation between predation by *Podisus nigrispi-nus* and developmental phase and density of its prey, *Plutella xylostella*. Entomologia Experimentalis et Applicata 2012;145(1): 30–37.

[44] Brito JP, Vacari AM, Thuler RT, De Bortoli SA. Aspectos biológicos de *Orius insidio-sus* (Say, 1832) predando ovos de *Plutella xylostella* (L., 1758) e *Anagasta kuehniella* (Zeller, 1879). Arquivos do Instituto Biológico 2009;76(4): 627–633.

[45] Kogan, M. Integrated pest management: historical perspectives and contemporary development. Annual Review of Entomology 1998;43: 243–270.

[46] Baek JH, Kim JL, Lee D, Chung BK, Miyata T, Lee SH. Identification and characteri-zation of ace1-type acetylcholinesterase likely associated with organophosphate re-sistance in *Plutella xylostella*. Pesticide Biochemistry and Physiology 2005;81(3): 164–175.

[47] Medeiros PT, Sone EH, Soares CMS, Dias JMCS, Monnerat RG. Avaliação de produ-tos à base de *Bacillus thuringiensis* no controle da traça-das-crucíferas. Horticultura Brasileira 2006;24(2): 245–248.

[48] Khan MFR, Griffin RP, Carner GR, Gorsuch CS. Susceptibility of Diamondback Moth, *Plutella xylostella* (L.) (Lepidoptera: Plutellidae), from collard fields in South Carolina to *Bacillus thuringiensis*. Journal of Agricultural and Urban Entomology 2005;22(1): 19–26.

[49] Thuler AMG, Thuler RT, Cícero ES, De Bortoli SA, Lemos MVF. Estudo da variabili-dade gênica em isolados brasileiros de *Bacillus thuringiensis* para emprego no con-trole biológico de *Plutella xylostella*. Boletín de Sanidad Vegetal Plagas 2007;33(3): 409–417.

[50] Baxter SW, Zhao J, Shelton AM, Vogel H, Heckel DG. Genetic mapping of Bt-toxin binding proteins in a Cry1A-toxin resistant strain of diamondback moth *Plutella xy-lostella*. Insect Biochemistry and Molecular Biology 2008;38(2): 125–135.

[51] Viana CLTP, De Bortoli SA, Thuler RT, Goulart RM, Thuler AMG, Lemos MVF, Fer-raudo AS. Efeito de novos isolados de *Bacillus thuringiensis* Berliner em *Plutella xylos-tella* (Linnaeus, 1758) (Lepidoptera: Plutellidae). Científica 2009;37(1): 22–31.

[52] Maeda M, Mizuki E, Nakamura Y, Hatano T, Ohba M. Recovery of *Bacillus thurin-giensis* from marine sediments of Japan. Current Microbiology 2000;40(6): 418–422.

[53] Martínez C, Caballero P. Contents of cry genes and insecticidal activity of *Bacillus thuringiensis* strains from terrestrial and aquatic habitats. Journal of Applied Microbi-ology 2002;92(4): 745–752.

[54] BRAVO, A.; GILLL, S. S.; SOBERÓN, M. Mode of action of *Bacillus thuringiensis* Cry and Cyt toxins and their potential for insect control. Toxicon 2007;49(4): 423–435.

[55] Estruch JJ, Warren GW, Mullins MA, NYE GJ, Craig JA. Vip3A, a novel *Bacillus thur-ingiensis* vegetative insecticidal protein with a wide spectrum of activities against

Lepidopteran insects. Proceedings of the National Academy of Science 1996;93(11): 5389–5394.

[56] Knowles BH, Ellar DJ. Colloid-osmotic lysis is a general feature of the mechanism of action of *Bacillus thuringiensis* δ-endotoxins with different insect specificity. Biochimica et Biophysica Acta 1987;924(3): 509–518.

[57] Kurouac M, Vachon V, Noel JF, Girard F, Schwartz JL, Laprade R. Amino acid and divalent ion permeability of the pores formed by the *Bacillus thuringiensis* toxins Cry1Aa and Cry1Ac in insect midgut brush border membrane vesicles. Biochimica et Biophysica Acta 2002;1591(2): 171–179.

[58] Vachon V, Laprade R, Schwartz JL. Current models of the mode of action of *Bacillus thuringiensis* insecticidal crystal proteins: a critical review. Journal of Invertebrate Pathology 2012;111(1): 1–12.

[59] Mohan M, Sushil SN, Selvakumar G, Bhatt JC, Gujar GT, Gupta HS. Differential toxicity of *Bacillus thuringiensis* strains and their crystal toxins against high-altitude Himalayan populations of diamondback moth, *Plutella xylostella* L. Pest Management Science 2009;65(1): 27–33.

[60] Crickmore N, Zeigler DR, Feitelson J, Schnepf E. van Rie J, Lereclus D, Baum J, Dean DH. Revision of the nomenclature for the *Bacillus thuringiensis* pesticidal crystal proteins. Microbiology and Molecular Biology Reviews 1998;62(3): 807–813.

[61] Silveira LFV, Polanczyk RA, Pratissoli D, Franco CR. Seleção de isolados de *Bacillus thuringiensis* Berliner para *Tetranychus urticae* Koch. Arquivos do Instituto Biológico 2011;78(2): 273-278.

[62] van Frankenhuyzen K. Insecticidal activity of *Bacillus thuringiensis* crystal proteins. Journal of Invertebrate Pathology 2009;101(1): 1–16.

[63] Ministério da Agricultura Pecuária e Abastecimento. Agrofit: sistema de agrotóxicos fitossanitários. http://agrofit.agricultura.gov.br/agrofit_cons/principal_agrofit_cons/ (accessed 15 August 2012).

[64] Maxwell EM, Fadamiro HY, Mclaughlin JR. Suppression of *Plutella xylostella* and *Trichoplusia ni* in cole crops with attracticide formulations. Journal of Economic Entomology 2006;99(4): 6–17.

[65] Monnerat RG, Leal-Bertioli SCM, Bertioli DJ, Butt TM, Bordat D. Caracterização de populações geograficamente distintas da traça-das-crucíferas por suscetibilidade ao *Bacillus thuringiensis* Berliner e RAPD-PCR. Horticultura Brasileira 2004;22(3): 607–609.

[66] Medeiros PT, Pereira MN, Martins ES, Gomes ACMM, Falcão R, Dias JMCS, Monnerat RG. Seleção e caracterização de estirpes de *Bacillus thuringiensis* efetivas no controle da traça-das-crucíferas *Plutella xylostella*. Pesquisa Agropecuária Brasileira 2005;40(11):1145–1148.

[67] Salama HS, Foda MS, El-Sharaby A, Matter M, Khalafallah M. Development of some lepidopterus cotton pests as affected by exposure to sublethal levels of endotoxins of *Bacillus thuringiensis* for different periods. Journal of Invertebrate Pathology 1981;38(2): 220–229.

[68] Polanczyk RA, Alves S. *Bacillus thuringiensis*: uma breve revisão. Agrociência 2003;7(2): 1–10.

[69] Silva LKF, Carvalho AG. Patogenicidade de *Bacillus thuringiensis* (Berliner, 1909) em lagartas de *Urbanus acawoios* (Williams, 1926) (Lepidoptera: Hesperiidae). Arquivos do Instituto Biológico 2004;71(2): 249–252.

[70] Baksh A, Khan A. Pathogenicity of *Paecilomyces tenuipes* to diamondback moth, *Plutella xylostella* at three temperatures in Trinidad. International Journal of Agriculture and Biology 2012;14(2): 261–265.

[71] Freed S, Jin FL, Naeem M, Ren SX, Hussian M. Toxicity of proteins secreted by Entomopathogenic fungi against *Plutella xylostella* (Lepidoptera: Plutellidae). International Journal of Agriculture and Biology 2012;14(2): 291–295.

[72] Xu D, Ali S, Huang Z. Insecticidal activity influence of 20-Hydroxyecdysone on the pathogenicity of *Isaria fumosorosea* against *Plutella xylostella*. Biological Control 2011;56(3): 239–244.

[73] Godonou I, James B, Atcha-Ahowé C, Vodouhe S, Kooyman C, Ahanchédé A, Korie S. Potential of *Beauveria bassiana* and *Metarhizium anisopliae* isolates from Benin to control *Plutella xylostella* (L.) (Lepidoptera: Plutellidae). Biological Control 2009;28(3): 220–224.

[74] Silva VCA, Barros R, Marques EJ, Torres JB. Suscetibilidade de *Plutella xylostella* (L.) (Lepidoptera: Plutellidae) aos fungos *Beauveria bassiana* (Bals.) Vuill. e *Metarhizium anisopliae* (Metsch.) Sorok. Neotropical Entomology 2003;32(4): 653–658.

[75] Brusselman E, Beck B, Pollet S, Temmerman F, Spanoghe P, Moens M, Nuyttens D. Effect of the spray application technique on the deposition of entomopathogenic nematodes in vegetables. Pest Management Science 2012;68(3): 444–453.

[76] Schroer S, Sulistyanto D, Ehlers RU. Control of *Plutella xylostella* using polymer-formulated *Steinernema carpocapsae* and *Bacillus thuringiensis* in cabbage fields. Journal of Applied Entomology 2005;129(4): 198–204.

[77] Schroer S, Ziermann D, Ehlers RU. Mode of action of a surfactant-polymer formulation to support performance of the entomopathogenic nematode *Steinernema carpocapsae* for control of diamondback moth larvae (*Plutella xylostella*). Biocontrol Science and Technology 2005;15(6): 601–613.

[78] Castelo Branco M, Melo CA. Resistência a abamectin e cartap em populações de traça-das-crucíferas. Horticultura Brasileira 2002;20(4): 541–543.

[79] Miles M. Insect Pest Management II – *Etiella*, False Wireworm and Diamondback Moth. GRDC Research updates. http://www.grdc.com.au, 2002 (accessed 20 August 2012).

[80] Micic S. Chemical Control of Insect and Allied Pests of Canola. Farmnote No. 1/2005. Department of Agriculture, South Perth, Western Australia, Australia; 2005.

[81] Carazo ER, Cartin VML, Monge AV, Lobo JAS, Araya LR. Resistencia de *Plutella xylostella* a deltametrina, metamidofós y cartap em Costa Rica. Manejo Integrado de Plagas 1999;53: 52–57.

[82] Castelo Branco M, França FH, Medeiros MA, Leal JGT. Uso de inseticidas para o controle da traça-do-tomateiro e da traça-das-crucíferas: um estudo de caso. Horticultura Brasileira 2001;19(1): 60–63.

[83] Thuler RT, De Bortoli SA, Barbosa JC. Eficácia de inseticidas químicos e produtos vegetais visando ao controle de *Plutella xylostella*. Científica 2007;35(2): 166–174.

[84] Georghiou G, Lagunes-Tejada A. The ocurrence of resistance to pesticides in arthropods. An index of cases reported through 1989. Rome: FAO; 1991. p318.

[85] Gallo D, Nakano O, Silveira Neto S, Carvalho RPL, Baptista GC, Berti Filho E, Parra JRP, Zucchi RA, Alves SB, Vendramim JD, Marchini LC, Lopes JRS, Omoto C. Entomologia Agrícola. Piracicaba: FEALQ; 2002. p920.

[86] Carballo VM, Hruska AJ. Períodos críticos de proteccion y efecto de la infestacion de *Plutella xylostella* L. (Lepidoptera: Plutellidae) sobre el rendimiento del repollo. Manejo Integrado de Plagas 1989;14: 46–60.

[87] Castelo Branco M, Villas Bôas GL, França FH. Nível de dano de traça-das-crucíferas em repolho. Horticultura Brasileira 1996;14(2): 154–157.

[88] Castelo Branco M, Gatehouse AG. Insecticide resistance in *Plutella xylostella* (Lepidoptera: Yponomeutidae) in the Federal District, Brazil. Anais da Sociedade Entomológica do Brasil 1997;26(1): 75–79.

[89] Ulmer BC, Gillot C, Woods D, Erlandson M. Diamondback moth, *Plutella xylostella* (L.), feeding and oviposition preferences on glossy and waxy *Brassica rapa* (L.) lines. Crop Protection 2002;21(4): 327–331.

[90] De Bortoli SA, Thuler RT, Lopes BS. Efeito de lufenuron e azadiractina sobre adultos de *Plutella xylostella* (Lepidoptera: Plutellidae). Científica 2006;34(1): 53–58.

[91] Lima MPL, Barros R. Toxicidade de lufenuron para lagartas de *Plutella xylostella* (L., 1758) (Lepidoptera: Plutellidae). Revista Ômega 2000;9(1): 52–54.

[92] O'Brien RD. Toxic phosphorus esters. New York: Academic; 1960. p434.

[93] Czepak C, Fernandes PM, Santana HG, Takatsuka FS, Rocha CL. Eficiência de inseticidas para o controle de *Plutella xylostella* (Lepidoptera: Plutellidae) na cultura do re-

polho (*Brassica oleraceae* var. *capitata*). Pesquisa Agropecuária Tropical 2005;35(2): 129–131.

[94] Bacci L, Picanço MC, Da Silva EM, Martins JC, Chediak M, Sena ME. Seletividade fisiológica de inseticidas aos inimigos naturais de *Plutella xylostella* (L.) (Lepidoptera: Plutellidae) em brássicas. Ciência e Agrotecnologia 2009;33: 2045–2051.

[95] Mordue AJ, Blackwell A. Azadirachtin: an update. Journal of Insect Physiology 1993;39(11): 903–924.

[96] Jesus FG, Paiva LA, Gonçalves VC, Marques MA, Boiça Júnior AL. Efeito de plantas inseticidas no comportamento e biologia de *Plutella xylostella* (Lepidoptera: Plutellidae). Arquivos do Instituto Biológico 2011;78(2): 279–285.

[97] Panda N, Khush GS. Host Plant Resistance to Insects. Wallingford, USA: CAB International; 1995, p448.

[98] Kessler A, Baldwin IT. Plant responses to insect herbivore: the emerging molecular analysis. Annual Review of Plant Biology 2002;53: 299–328.

[99] Eigenbrode SD, Pillai SK. Neonate *Plutella xylostella* responses to *surfasse* wax componentes of a resistant cabbage (*Brassica oleraceae*). Journal of Chemical Ecology 1998;24(10): 1611–1627.

[100] Rask L, Andréasson E, Ekbom B, Eriksson S, Pontoppidan B, Meijer J. Myrosinase: gene family evolution and herbivore defense in Brassicaceae. Plant Molecular Biology 2000;42(1): 93–113.

[101] Thuler RT, De Bortoli SA, Hoffmann-Campo CB. Classificação de cultivares de brássicas com relação a resistência a traça-das-crucíferas e a presença de glucosinolatos. Pesquisa Agropecuária Brasileira 2007;42(4): 467–474.

[102] Boiça Junior AL, Tagliari SRA, Pitta RM, Jesus FG, Braz LT. Influência de genótipos de couve (*Brassica oleracea* L. var. *acephala* DC.) na biologia de *Plutella xylostella* (L., 1758) (Lepidoptera: Plutellidae). Ciência e Agrotecnologia 2011;35(4): 710–717.

[103] Freitas LM, Junqueira AMR, Michereff Filho M. Potencial de uso do silício no manejo integrado da traça-das-crucíferas, *Plutella xylostella*, em plantas de repolho. Revista Caatinga 2012;25(1): 8–13.

[104] Gu H, Fitt GP, Baker GH. Invertebrate pests of canola and their management in Australia: a review. Australian Journal of Entomology 2007;46(3): 231–243.

[105] Cai HJ, Li SY, Ryall K, You MS, Lin S. Effects of intercropping of garlic or lettuce with Chinese cabbage on the development of larvae and pupae of diamondback moth (*Plutella xylostella*). African Journal of Agricultural Research 2011;6(15): 3609–3615.

[106] Mchugh JJ, Foster RE. Reduction of diamondback moth (Lepidoptera: Plutellidae) infestation in head cabbage by overhead irrigation. Journal of Economic Entomology 1995;88(1): 162–168.

[107] Moldrup ME, Geu-Flores F, De Vos M, Olsen CE, Sun J, Jander G, Halkier BA. Engineering of benzylglucosinolate in tobacco provides proof-of-concept for dead-end trap crops genetically modified to attract *Plutella xylostella* (diamondback moth). Plant Biotechnology Journal 2012;10(4): 435–442.

[108] Broad ST, Lisson SN, Mendham NJ. Agronomic and gross margin analysis of an insect pest suppressive broccoli cropping system. Agricultural Systems 2009;102(1–3): 41–47.

[109] Broad ST, Schellhorn NA, Lisson SN, Mendham NJ. Host location and oviposition of lepidopteran herbivores in diversified broccoli cropping systems. Agricultural and Forest Entomology 2008;10(2): 157–165.

[110] Zong GH, Yan SQ, Liang XM, Wang DQ, Zhang JJ. Synthesis of the sex pheromone of *Plutella xylostella* (L.). Chinese Journal of Organic Chemistry 2011;31(12): 2126–2130.

[111] Nofemela RS. The ability of synthetic sex pheromone traps to forecast *Plutella xylostella* infestations depends on survival of immature stages. Entomologia Experimentalis et Applicata 2010;136(3): 281–289.

[112] Hopkinson RF, Soroka JJ. Air trajectory model applied to an in-depth diagnosis of potential diamondback moth infestations on the Canadial Prairies. Agricultural and Forest Meteorology 2010;150(1): 1–11.

[113] Evenden ML, Gries R. Assessment of commercially available pheromone lures for monitoring diamondback moth (Lepidoptera: Plutellidae) in Canola. Journal of Economic Entomology 2010;103(3): 654–661.

[114] Imenes SDL, Campos TB, Rodrigues Netto SM, Bergmann EC. Avaliação da atratividade de feromônio sexual sintético da traça-das-crucíferas, *Plutella xylostella* (L.) (Lepidoptera: Plutellidae), em cultivo orgânico de repolho. Arquivos do Instituto Biológico 2002;69(1): 81–84.

[115] Ahuja I, Rohloff J, Bones AM. Defence mechanisms of Brassicaceae: implications for plant-insect interactions and potential for integrated pest management. A review. Agronomy for Sustainable Development 2010;30(2): 311–348.

Chapter 4

Biological Control of Root Pathogens by Plant- Growth Promoting *Bacillus* spp.

Hernández F.D. Castillo, Castillo F. Reyes,
Gallegos G. Morales, Rodríguez R. Herrera and
C. Aguilar

Additional information is available at the end of the chapter

1. Introduction

At the present time, among the most important factors limiting production of different crops are soil-borne plant pathogens [1]. Which include the genera *Pythium, Rhizoctonia, Fusarium, Verticillium, Phytophthora spp, Sclerotinia, Sclerotium,* and *Rosellinia* [2]. By this reason, different methods have been used to control these pathogens [3]. Cultural practices and chemical control using synthetic fungicides are the most used control methods [4], however, use of some of these synthetic products has caused various problems due to environmental pollution, with consequences such as toxicity to humans, as well as resistance of certain pathogens to these fungicides [5]. An alternative to reduce the effect of these plant pathogens is the use of antagonistic microorganisms such as: some species of the genus*Bacillus* which is recognized as one of the most effective biological control agent because of their properties on pathogens growth inhibition [6-7]. Biological control has many advantages as an alternative in the integrated management of diseases such as little or no harmful side effects, rare cases of resistance, long-term control, completely or substantially eliminates the use of synthetic pesticides, cost / benefit ratio very favorable; prevents secondary diseases, not symptoms of poisoning and can be used as part of integrated disease management [8]. Generally, the mode of action of *Bacillus* is antibiosis by producing extracellular hydrolytic enzymes which decompose polysaccharides, nucleic acids, other way are: production of antibiotics such as bacitracin, polymyxin, and gramicidin, [9-11], competition to occupy an ecological niche and metabolize root exudates on pathogens affecting their growth [12-13]. Also, activating plant resistance induction when installed in the roots and leaves which induces plant to produce phytoalexins which give resistance against attack by fungi, bacteria and pathogenic nematodes [14], reducing in these ways disease incidence.

2. Overview of Bacillus

The genus *Bacillus* Cohn was established in 1872, initially with two prominent forming endospores species: *Bacillus anthracis* and *B. subtilis*, actually, this genus has suffered considerable taxonomic changes, until early 1900, taxonomists not only restricted the genus to endospore forming bacteria, having that the number of species assigned to this genus were 146 in the 5th edition of Bergey's Manual of Systematic Bacteriology. Subsequent comparison studies by Smith et al. and Gordon et al. over 1114 strains of aerobic bacteria forming endospores (PGPR) helped to reduce this number to 22 well-defined species same as reported in the 8th edition of Bergey's Manual of Systematic Bacteriology [15]. Bacillales is the order to which Bacillaceae family belongs within the genus *Bacillus*. This genus is characterized by having a rod shape within the group of Gram positive [16-17], and is therefore classified as strict aerobes or facultative anaerobes [18] and integrated by 88 species [15]. A feature associated with this genus is that it forms a type of cell called endospore as a response to adverse growth conditions which distorts the structure of the cell. This spore form is resistant to high temperatures and current chemical disinfectants [19]. This genus is abundant in various ecological niches which include soil, water and air [18, 20], it is also found as food contaminants. Generally, *Bacillus* species used in bio-control are mobile, with peritrichous flagella, but yet some species are of interest in human medicine (*B. anthracis*) which are characterized as being stationary [21].

2.1. Ecology and habits

Distribution and habitat of *Bacillus* are very diverse; some species have been isolated from soil micro-flora adjacent to plants rhizosphere, water, air and food as contaminants [18, 20]. Eco-physiological criteria commonly used to group different species such as vertebrate pathogens, insect pathogens, antibiotics producer, nitrogen-fixing, denitrifying, thermophilic, psychotropic halophilic, alkali and acidifies rows. For example *B. thuringiensis* is considered an insect pathogen and is used as a bio-pesticide, it has been isolated from soil, and abundantly found worldwide [22-24], in soil remains largely in the form of endospores [25], particles of dust in suspension [26], insect bodies sick or dead [27], also is found in stored products [28-29], food [30], marine sediments [32], and even as opportunistic human pathogen [33]. Furthermore, *Bacillus* species are found abundantly in plant leaves [34-38]. In conclusion, the *Bacillus* genus has a cosmopolitan distribution (Table 1).

Organism	Reference strain isolated from	Common habitats and comments
B. acidocaldarius	Hot springs	Acid hot springs and soils, enrichment from neutral soils have failed.
B. alcalophilus	Human feces	soil, water, dung
B. alvei	Honeybee larvae suffering from European foulbrood	Soil, this specie is a saprophyte but common in bees with European foulbrood
B. aminovorans	Soil	
B. amyloliquefaciens	Soil	soil, industrial amylase fermentations
B. amylolyticus	Soil	
B. aneurinolyticus	human feces	
B. azotofixans	Soil	soil, rhizosphere of various grasses
B. azotoformans	Soil	Soil
B. apiaries	Dead larvae of honeybee	
B. badius	Human feces	Dust, coastal waters, soil
B. benzoevarans	Soil	Soil
B. brevis	Soil	Foods, soil, seawater, and sediments
B. cereus	Soil	soil, foods, especially dried foods, spices, and milk; seawater and sediments
B. circulans	Soil	Widespread in soil and decomposing vegetables; medicated creams, Relatively scarce in soil.
B. cirroflagellosus	marine mud	
B. coagulans	evaporated milk	Beet sugar, canned foods, especially vegetables; medicated creams, relatively scarce in soil.
B. epiphytus	marine phytoplankton	
B. fastidiosus	Soil	soil, poultry litter
B. firmus	Soil	soil, seawater and marine sediments, salt marshes
B. freudenreichii		Soil, river water, and sewage
B. globisporus	Soil	soil, mud, and water
B. insolitus	Soil	soil, mud, water and frozen foods
B. laevolacticus	Rhizosphere of ditch crowfoot	rhizosphere of plants
B. larvae	honeybee larvae suffering from American foulbrood	Infected brood and honey combs. Presumable in soil around hives of bees
B. laterosporas	Soil	soil, water, dead honeybee larvae, rumen of animals
B. lentimorbus	hemolymph of larvae of Japanese beetle	causes milky disease of scarabaeidae larvae

Organism	Reference strain isolated from	Common habitats and comments
B. lentus	Soil	Seawater, marine sediments, salt marshes and soil. Spices including black and red pepper
B. lecheniformis	Soil	soil, marine and freshwaters; foods, particularly dried foods, spices and cocoa beans, compost, rumen of cattle
B. macerans	unknown	foods and vegetables, compost
B. macquariensis	soil from Macquarie island	Unknown
B. macroides	cow dung	decaying material
B. marinus	seawater	Unknown
B. megaterium	Soil	soil including desert soil, seawater and marine sediments, cocoa bean, dried foods and spices
B. pacificus	sand from seashore	Seawater
B. pantothenticus	Soil	generally considered to be a soil inhabitant but also isolated from pharmaceutical products
B. pasteutii	Soil	soil, water, sewage, urinals
B. polymyxa	Soil	widely distributed in soil, decomposing plant matter and water
B. popilliae	Commercial spore dust	causes milky disease of scarabaeidae larvae
B. psychrophilus	Soil	soil, water, mud, frozen foods vegetables
B. pulvifaciens	dead larvae of honeybee	Unknown
B. pumilus	Soil	Ubiquitous in soil. Also found in seawater and marine sediments. Common in dried foods.
B. racemilactius	rhizosphere of wild lettuce	rhizosphere of plants
B. schlegelii	sediments of eutrophic lake	Unknown
B. sphaericus	Soil	soil, marine and freshwaters sediments and foods
B. stearothermoplilus	unknown	soil, foods including milk, canned foods and sugar beet, dried foods
B. subtilis	Soil	soil, marine and freshwater and sediments, foods including spices, cocoa, pulses, seeds and bread
B. thermoglucosidasius	Soil	Unknown
B. thiaminolyticus	Human feces	Unknown
B. thuringiensis		Pathogenic for lepidopteran larvae, common in soil.
B. xerothermodurans	Soil	Unknown

Table 1. Sources and common habitats of aerobic endospore forming bacteria of *Bacillus* genus, [39].

2.2. Importance how antifungal agents

Many species of *Bacillus* including *B. subtilis*, *B. licheniformis*, *B. pumilus*, *B. amyloliquefaciens*, *B. cereus*, *B. mycoides* and *B. thuringiensis*, are known to suppress growth of several fungal pathogens such as *Rhizoctonia, Fusarium, Sclerotinia, Sclerotium, Gaeummanomyces, Nectria, Pythium, Phytophthora* and *Verticillium* [20, 40-43]. The main property of antagonist bacterial strains is production of antifungal antibiotics [44-45], which seem to play a major role in biological control of plant pathogens [6, 44, 46-49] and post-harvest spoilage fungi [50]. Many of these antifungal substances have been characterized and identified as peptide antibiotics [51]. Antifungal peptides produced by *Bacillus* species: iturins [20, 52-53] are: mycosubtilins [54-55], bacillomycins [56-57], surfactins [58-59], fungistatins [60-61], and subsporins [62-63]. Most of these antibiotics are cyclic peptides composed entirely of amino acids, but some may contain other residues. However, a few antibiotic peptides are linear such as rhizocticins [64]. *Bacillus* spp. produces also a range of other metabolites including chitinases and other cell wall-degrading enzymes [65-68], and volatiles compounds [68-70] which elicit plant resistance mechanisms [14, 71].

The amount of antibiotics produced by bacilli class was approaching 167 [45], being 66 derived from *B. subtilis*, 23 from *B. brevis* and the remaining antibiotic peptides are produced by other species of *Bacillus*. The main antibiotic producers of this genus are *B. brevis* (gramicidin, tyrothricin) [72], *B. licheniformis* (bacitracin), *B. polymyxa* (polymyxin, colistin), *B. pumilus* (pumulin), *B. subtilis* (polymyxin, difficidin, subtilin, mycobacillin, bacitracin), *B. cereus* (cerexin, zwittermicin), *B. circulans* (circulin), *B. laterosporus* (laterosporin) [14, 68-71].

2.3. Collection and isolation of *Bacillus*

Traditional tools for determining composition of the soil bacterial community and diversity are based largely on *in vitro* culture methods. Typically, solid organic medium is inoculated with dilutions of a suspension of soil, then incubated and the colonies obtained are purified further sub culturing into another medium [73]. Heat treatment or pasteurization is the most used technique for selecting spores. These techniques are very powerful because they are selective to remove all non-spore forming microorganisms, and are very efficient for obtaining populations of bacteria from spores, recommended temperatures oscillate between 65 to 70 ° C for 15 minutes [74-75]. However, heat treatment has to be adapted to certain species because endospores of some strains of bacteria are more resistant to heat than others, while incubation time used can vary from 3 to 30 min [76]. It is recommended to start heating at a relatively low temperature (70 or 75 ° C) and gradually increasing to achieve an optimum temperature [77]. To isolate endospores, some authors have taken advantage of spore tolerance to diverse stress conditions, for example, Koransky et al. [78] concluded that treatment with ethanol (50%) for 1 h is an effective technique to selectively isolate spore- forming bacteria, as effective as heat treatment to 80 ° C for 15 minutes. Patel et al. [79] confirmed this finding by isolating *Bacillus* strains from food residues, both by heating at 65 ° C for 45 minutes and incubation with ethanol. Soil drying may also be used as a selective method to isolation by striking desiccation tolerance of spores, which can therefore survive for long periods of time under these conditions. Drying treatment is probably more gentle that heating or ethanol incubation. Eman et al. [80] reported that vegetative cells were killed by addition of chloroform (1% v / v) however; this technique has not been validated. An interesting selection process, which is different from classical heat treatment was developed by Travers et al. [81] for isolation of *Bacillus thuringiensis*, which makes use of ethyl (ethyl selection), *B. thuringiensis* is selectively inhibited by sodium acetate (0.25 M), while most unwanted spore-forming species allowed to germinate. Then all non-sporulating bacteria were eliminated by heat treatment at 80 ° C for 3 min. Subsequently, surviving spores are germinated on enriched agar medium. Even if some other species of *Bacillus* are also selected by this method, such as *B. sphaericus* and *B. cereus*, this technique is commonly used for studying the diversity worldwide of *B. thuringiensis* [22, 77]. A modification to the method promotes greater sporulation spore production by stimulating shock before applying stress. For example, some authors suggest suspending one gram of soil in 50 mL of sporulation medium after incubation at 37 ° C under stirring for 48 hours before killing vegetative cells by heat treatment [80], while others proposed in soil suspensions incubate the culture broth at different temperatures for 5 days to allow better maturation of spores [77].

2.4. Biochemical identification

Biochemical test were the traditional method for bacteria identification to specie level, after that, strains are located at the genus taxonomically, based on characteristics of colony growth in artificial medium, form cell unit, presence, number and orientation of locomotive units, Gram stain, spore form and specific environmental conditions of growth and finally the specific use of carbon sources (biochemical tests) gave its metabolic diversity (Table 2 and 3).

	B. amylolique-faciens	*B. pumilus*	*B. subtilis*	*B. licheniformis*	*B. thuringiensis*	*B. cereus*	*B. mycoides*	*B. fastidiosus*	*B. firmus*	*B. lentus*	*B. megaterium*	*B. bodius*	*B. antracis*
Cell diameter"/>1.0 um	-	-	-	-	+	+	+	+	-	-	+	-	+
Parasporal crystals	-	-	-	-	d	-	-	-	-	-	-	-	-
Anaerobic growth	-	-	-	+	+	+	+	-	-	-	-	-	+
Voges Proskauer test	+	-	+	+	d	+	+	NG	-	-	-	-	+
egg yolk lecithin's	-	-	-	-	+	+	+	-	-	-	-	-	+
growth in lysozyme	-	d	d	d	+	+	+	ND	-	-	-	-	+
Acid from													
d-glucose	+	+	+	+	+	+	+	NG	+	+	+	-	+

	B. amyloliquefaciens	*B. pumilus*	*B. subtilis*	*B. licheniformis*	*B. thuringiensis*	*B. cereus*	*B. mycoides*	*B. fastidiosus*	*B. firmus*	*B. lentus*	*B. megaterium*	*B. bodius*	*B. antracis*
l- arabinose	d	+	+	+	-	-	-	NG	-	+	d	-	-
d-xylose	D	+	+	+	-	-	-	NG	-	+	d	-	-
d-mannitol	+	+	+	+	-	-	-	NG	+	+	d	-	-
hydrolysis of													
Starch	+	-	+	+	+	+	+	-	+	+	+	-	+
Casein	+	+	+	+	+	+	+	-	+	d	+	+	+
nitrate reduction	+	-	+	+	+	+	+	-	d	d	d	-	+
degradation of tyrosine	-	-	-	-	ND	+	ND	-	d	-	d	+	d
Growth in 7% NaCl	+	+	+	+	+	d	d	-	+	d	d	ND	+
Growth at													
10°C	ND	+	d	-	d	d	d	+	d	ND	+	-	-
50°C	d	d	d	+	-	-	-	-	-	-	-	+	-
55°C	ND	-	-	+	-	-	-	-	-	-	-	-	-
Utilization of													
Citrate	d	+	+	+	+	+	d	-	-	-	+	-	D
Propionate	ND	-	-	+	ND	ND	ND	-	-	-	ND	-	ND

Table 2. Differential characteristics of *Bacillus* species with ellipsoidal spores (Group I), [39]. += 90 or more of strains positive catalase; - = 10 or more of strains negative catalese; d= substancial proportion of specie differ; ND= Not done; NG= no growth.

2.5. Molecular identification

Bacillus species with diverse physiological traits require development of biochemical tests for identification [82]. But advances in chromatographic analysis using whole cell fatty acid methyl esters (FAME) profiles allows doing this technique sufficiently sensitive and reliable for grouping *Bacillus* to specie level [83-84]. Identification has become even more sensitive, by analysis of ribosomal DNA regions (16S rDNA) sequencing [85-87], and sequence analysis of gyrase B (gyrB) which has proved immensely valuable information for phylogenetic analysis of bacteria [88-90]. Using 16S rDNA sequence, have been identified 5 groups within the genus *Bacillus*, where group 1 (group *B subitilis*) comprises species *B. amyloliquefaciens*, *B. subtilis*, *B. pumilus* and *B. licheniformis* [9, 39, 91].

Test	Bs[y]	Ba	Bl	B1	B3	B9	B13
Gram staining	+[z]	+	+	+	+	+	+
Flagella staining	+	+	+	+	+	+	+
RYU Test	-	-	-	-	-	-	-
Oxidase	-	-	-	-	-	-	-
Catalase	+	+	+	+	+	+	+
Oxidation	+	+	-	+	-	+	+
Fermentation	+	+	+	+	+	+	+
Motility	+	+	+	+	+	+	+
Spore Posicion							
Terminal	-	-	-	-	-	-	-
Central	+	+	+	+	+	+	+
Subterminal	-	-	-	-	-	-	-
Colony Growth:							
45ºC	+	+	+	+	+	+	+
65ºC	-	+	-	+	-	-	-
pH Growth at 5.7	+	+	+	+	+	+	+
NaCl Growth:							
7%	+	+	+	+	+	+	+
5%	+	-	+	-	+	+	+
3%	+	-	+	-	+	+	+
citrate utilization	+	+	+	+	+	+	+
Anaerobic growth in glucoseglucose	-	+	+	+	+	+	+
Glucose							
Acidic Forms:							
Arabinose	+	+	+	+	+	+	+
Manitol	+	-	+	-	+	+	+
Xylose	+	+	+	nd	nd	nd	nd
Voges-Proskauer	+	+	+	+	+	+	+
Hydrolysis starch	+	+	+	+	+	+	+

Table 3. Results on identification of *Bacillus* isolates B1, B3, B9 and B13 by biochemical tests, [9], and Bs = *Bacillus subtilis*, Ba = *B. amyloliquefaciens*; Bl = *B. licheniformis*. Positive test Z = +; negative = -, nd = not determined. [9].

2.6. Antifungal effect *in vitro*, greenhouse and field

Bacillus species have been reported also as growth promoters of certain crops [92], and with antifungal properties, for example *B. amyloliquefaciens* has been reported as a specie with antifungal activity against *Colletotrichum dematium, Colletotrichum lagenarium, Rosellinia necatrix, Pyricularia oryzae, Agrobacterium tumefaciens, Xanthomonas campestris* pv. *campestris* and *Xanthomonas campestris* pv. *vesicatoria in vitro* and *in vivo* [93-95]; antagonistic to *Botrytis elliptica*, under greenhouse conditions [96]; antagonistic to *Botrytis cinerea* in postharvest [97], in the biological control of *Rhizoctonia solani, Fusarium* spp. and *Pythium* spp. [98], as well as inductor of resistance mechanisms in plants [99]. *Bacillus licheniformis* is reported as a fungicide against a variety of pathogens, both as a preventive and curative particularly leaf spots and blights, and a growth-promoting bacteria with production likely gibberellins [100]. *Bacillus subtilis* is the most studied and has been reported as growth promoter and antagonistic to a variety of pathogens such as *Phytophthora cactorum, Sclerotium cepivorum, Fusarium oxysporum, Rhizoctonia solani, Alternaria carthami, Phytophthora capsici,* and *Fusarium solani* among others, in different cultures and evaluated *in vitro*, greenhouse and field level [101-103], so that *Bacillus* strains can be using as an alternative in biological control for management plant disease.

2.6.1. In vitro studies

Results of *in vitro* research using *Bacillus* spp. as biocontrol agent against various soil pathogens, have reported positive responses through observing a negative effect on pathogen growth (Figure 1), per example against *Alternaria dauci* and *Rhizoctonia solani*, foliage and soil pathogens, respectively. In the Table 4 and 5, are showed some effect on pathogen mycelia inhibition by action of *Bacillus*, up to 50% compared to treatment control. Furthermore, in the case of *A. dauci*, greater control was observed with biocontrol agents compared to chemical treatment.

Treatments	Mycelia Inhibition (%)
Strain *Bacillus* B3*	35.55 a
Strain *Bacillus* B9	40.44 ab
Strain *Bacillus* B15	29.44 ab
L. tridentata extract (4000 ppm)	22.22 b
L. tridentata extract (2000 ppm)	11.11 c
Witness [1]	0 d

Table 4. *In vitro* mycelia inhibition of *Rhizoctonia solani* with *Bacillus* spp. strains and *Larrea tridentata* extract. * Summated dozes of *Bacillus* strains were 1x10^6 cfu / ml, [1] without agrochemicals, [46]. Values in the same column followed by different letters are significantly at p <0.05.

Treatments	Mycelia inhibitions (%)
Strain *Bacillus* B1*	53.44[a]
Strain *Bacillus* B3	48.44b
Strain *Bacillus* B9	40.31c
Strain *Bacillus* B13	46.25b
Strain *Bacillus* B15	0f
Strains *Bacillus* Mix	0f
Q-L 2000-2000 ppm	14.06d
Q-L 2000-1000	4.06e
Q-L 1000-2000	1.88ef
Q-L 1000-1000	0f
Witness[1]	0f

Table 5. *In vitro* mycelia growth inhibition of *Alternaria dauci* by *Bacillus* spp. and chitosan-Larrea (Q-L) suspensions. * Strains of *Bacillus subtilis*. Values in the same column followed by different letters are significantly at p <0.05, [44].

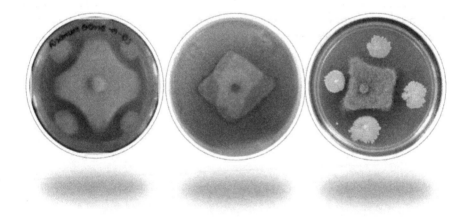

Figure 1. Effect *B. subtilis* in inhibition of mycelia growth of *Fusarium sp., Alternaria dauci* and *Rhizoctonia solani*.

2.6.2. Greenhouse studies

Results under greenhouse conditions, present good evidence of *Bacillus* as biocontrol source for pathogens involved in diseases of root and plant foliage, to cause a decrease in disease development in both incidence and severity. In table 6 is showed that application of *Bacillus* on carrot foliage allowed a control of *A. dauci* incidence up to 25%, which represents a control to 2 times more than the chemical treatment used for its control.

	Incidence	Severity
Treatments	**%**	
Strain *Bacillus* B1	25d	0.5 de
Strain *Bacillus* B3	0e	0 e
Strain *Bacillus* B9	25d	0.5 de
Strain *Bacillus* B13	25d	0.5 de
Strain *Bacillus* B15	50c	1 d
Strains *Bacillus* Mix	50c	1 d
Q-L 2000-2000 ppm	50c	3 c
Fungicides synthetics Mix*	75b	4.24 b
Witness	100[a]	6.75 a

Table 6. Product effect of *Bacillus* based biological products and chemicals on incidence and severity of *Alternaria dauci* on carrot plants under greenhouse conditions. * Chlorothalonil, iprodione, propiconazole, thiabendazole and fluazinam, [44].

Likewise *Bacillus* use has favored not only reduction of symptoms and therefore incidence, but also helps to promote plant growth, which is expressed in greater plant height, as shown in Figure 2, there is an increase in tomato plants height by effect of a microcapsule formulation of *Bacillus* applied in the management of disease caused by *F. oxysporum* and *R. solani*, in contrast to the use of synthetic chemicals [104].

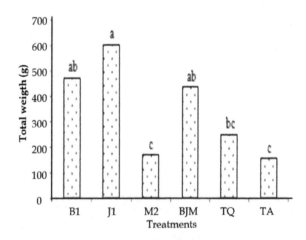

Figure 2. Average fruit yield of tomato plants cv. Floradade under greenhouse conditions subjected to different treatments with microcapsules containing strains of *Bacillus subtilis*, a chemical control (TQ) and a blank (TA).

2.6.3. Field experience

Most research has been conducted in laboratory or greenhouse, and virtually no field-level assessments have been reported. A study carried out using *Bacillus* spp. for control of diseases caused by soil fungi including: *F. oxysporum, R. solani* and *P. capsici* in pepper and tomato crops allowed control of diseases incidence as seen in Table 7, use of *Bacillus* at transplanting can reduce disease incidence in contrast to traditional treatments (fungicide application) as Folpat, Captan, Mancozeb much as 64% and 72% compared to untreated control with the most efficient strain, only 36 and 40 respectively with less efficient biological treatment.

	Harvest 1	B/TT	B/T	Harvest 2	B/TT	B/T	Harvest 3	B/TT	B/T
Treatments	Incidence %			Incidence %			Incidence %		
Bacillus B1	2.71 a	36	33	10.87 c	33	19	19.59 c	34	20
Bacillus B2	2.71 a	36	33	13.04 c	39	23	28.80 c	50	30
Bacillus B9	3.80 a	50	47	11.41 c	34	20	20.65 c	36	21
Bacillus B13	4.89 a	64	60	11.41 c	34	20	25.54 c	44	26
Bacillus Mix	3.25 a	43	40	15.76 c	48	28	20.11 c	35	21
Traditional treatment	7.61 a	100	93	33.13 b	100	59	57.61 b	100	60
Control	8.15 a	107	100	55.97 a	169	100	96.74 a	168	100

Table 7. Effect of *Bacillus* and commercial products on root rot incidence in *Capsicum annum* at different harvest times, [9]. Variance analysis used transformed data by arcsine. B/TT= *Bacillus* vs traditional treatments; B/T = *Bacillus* vs Control [9].

Furthermore, the suppressive effect was maintained over time or among harvest times, this indicates that *Bacillus* strains suppressed disease caused by soil fungi and maintained their remedial effect through harvest times as seen in Table 8 where disease incidence and severity was lower than that offered by the traditional treatment performed by the farmer.

Treatments	Incidence	Severity
	%	
Strain *Bacillus* B1	2.1 c	2.35 b
Strain *Bacillus* B3	3.05 ac	3.10 ab
Strain *Bacillus* B9	3.00 ac	3.05 ab
Strain *Bacillus* B13	2.75 bc	2.85 b
Strains *Bacillus* mix	2.90 ac	3.00 ab
Treatments Fungicides	3.5 ab	3.25 ab
Witness	3.85 a	3.85 a

Table 8. Effect of four strains of *Bacillus* and commercial products on severity of wilt and root rot by *Fusarium* spp., *Rhizoctonia solani* and *Phytophthora capsici*, on pepper (*Capsicum annum*) using scales to wilt and root rot, [9].

In the case of tomato same behavior was observed for disease development with respect to the presence of *Bacillus*, Table 9.

Treatments	Incidence (%)	Severity
Bacillus B1	0.0 d	0.0 c
Bacillus J1	0.0 d	0.0 c
Bacillus M2	12.0.c	1.5 c
B1J1M2 Mix	0.0 b	0.0 c
QT*	27.0 b	3.5 b
AT**	75.0 a	5.0 a
CV (%)	10.4	1.2

Table 9. Disease incidence and severity at harvest time of tomato plants cv. Florade subjected to different treatments with microcapsules containing strains of *Bacillus subtilis*. *Chemical control treatments; ** Absolute control treatment, Values with same letters area not statistically different (Tukey, p<0.01). [104].

The application in field of *Bacillus* sp on melon crops (Figure 3) for the management of disease caused by *F. oxysporum*, showed an effect in reducing disease incidence in 41% compared to the conventional chemical treatment (TA), and increases in yields of 26.5% higher than TA, however there were no significant differences in the brix degrees, consistency of fruit, but an increase in 12% in the number of fruits and 20% in the length guide, leaves and stem diameter was observed [105].

Figure 3. Treatments effect with *Bacillus subtilis* in fields on melon crops with high incidence of *Fusarium oxysporum*.

2.7. Effect on plant development and growth

The effects obtained by applying *Bacillus* as fungicide were positive for crop development, because *Bacillus* stimulated biomass production, increased number of flowers and fruiting mooring, as seen in Table 10, where *Bacillus* was applied, plants had an increase in height, flowering and fruiting compared to traditional crop management through use of synthetic agrochemicals. It is noteworthy that use only at the time of transplantation *Bacillus* and 20 days after the second application, which is manifested by loss of effect at 84 days, but yet still persists pathogen control, before such situation should be applied in successive moments to keep a *Bacillus* greater crop protection coverage.

				56 days	84 days	
	Height	Flowers	Fruits	Height	Flowers	Fruits Frutos
Treatments	**(cm)**	**(No.)**	**(No.)**	**(cm)**	**(No.)**	**(No.)**
Bacillus B1	39.95 ab	3.60 a	1.25 a	61.20 ab	13.35 a	6.00 a
Bacillus B3	42.75 a	3.60 a	1.05 a	60.05 ab	10.70 a	5.05 a
Bacillus B9	41.20 ab	3.05 a	1.05 a	59.40 ab	9.59 a	4.50 a
Bacillus B13	43.75 a	3.50 a	1.35 a	66.40 a	14.35 a	4.90 a
Bacillus mix	40.90 ab	3.15 a	1.40 a	63.55 ab	12.10 a	6.00 a
Traditional treatments	35.30 ab	2.75 a	0.60 a	58.10 ab	10.45 a	5.95 a
Control	32.80 b	2.65 a	0.55 a	55.55 b	12.00 a	5.00 a

Table 10. Effect of *Bacillus* strains on pepper (*Capsicum annuum*) cv. Caballero development, at 56 and 84 days after inoculation under field conditions. [9].

In a similar way, positive effects of *Bacillus* application were observed on crop yield, by prevent soil pathogen attack. In Table 11, is showed that application of *Bacillus* increased pepper yields in contrast to the traditional crop management by up to 74% cumulative assessment in three harvest times.

Treatment	Cut 1	B/TT	Cut 2	B/TT	Cut 3	B/TT	Yield (kg)	B/TT
	(kg)	**(%)**	**(kg)**	**(%)**	**(kg)**	**(%)**		**(%)**
BacillusB1	4.38 a	100	4.01 a	150	6.69 a	421	15.10 a	174
BacillusB3	3.32 ab	76	3.05 ab	114	4.01 b	252	10.39 b	120
Bacillus B9	2.73 ab	62	2.40 ab	90	4.04 b	254	9.17 b	106
Bacillus B13	3.79 ab	86	3.41 ab	127	4.01 b	252	11.16 ab	129
Bacillus mix	3.14 ab	72	2.80 ab	104	4.30 b	270	10.25 b	118
Traditional Treatments	4.39 a	100	2.68 ab	100	1.59 c	100	8.67 b	100
Control	1.79 b	41	1.71 b	64	0.57 c	36	4.08 c	47

Table 11. Effect of *Bacillus* strains on pepper (*Capsicum annuum*) cv. Caballero development at 99, 113 and 146 days after inoculation under field conditions. B / TT = *Bacillus* vs. traditional treatment, [9].

Bacillus favors growth of different plant parts such as stems and leaf area (Tables 12 and 13, Figure 4).

Figure 4. Effect of *Bacillussubtilis* in the biomass production of pepper plants (root). Treatments *Bacillus* and treatments without *Bacillus.*

Treatments	Height (cm)	Leaf area (cm2)
Bacillus B1	119.47 a	6857.01 b
Bacillus J1	118.65 a	7762.92 a
Bacillus M2	102.95 b	5393.32 b
Mixture B1J1M2	121.05 a	7022.90 a
*TQ	99.05 b	4007.51 c
**TA	98.9 b	4302.63 bc
CV (%)	3.11	9.29

Table 12. Height and leaf area of tomato plants cv. Floradade subjected to different treatments with microencapsulated strains of *Bacillus subtilis.* *: Chemical control treatment; **Absolute control treatment, Values with same letters are not statistically different (Tukey, p≤0.01), [104].

Treatments	Leaves	Stems	Roots
Bacillus B1	116.41 a	30.31 ab	31.28 c
Bacillus J1	107.57 a	31.92 ab	38.76 b
Bacillus M2	87.67 b	27.90 b	32.04 c
Mixture B1J1M2	94.51 b	34.69 a	96.63 a
*TQ	1 c	28.06 b	33.38 bc
**TA	26.21 d	14.57 c	13.92 d
CV (%)	6.54	8.32	6.97

Table 13. Biomass (g) production of tomato plants cv. Floradade subjected to different treatments with microcapsules containing strains of Bacillus subtilis. *Chemical control treatment; **: Absolute control treatment. Values with same letters are not statistically different (Tukey, p≤0.01), [104].

3. Conclusions

Use and application of biological control agents, such as *Bacillus* spp. prevents negative effects of pathogen attack on crops, providing an attractive option for sustainable agriculture due to their stimulating effects on plant growth, biomass production and its potential to increase plant production. In this chapter are mentioned clear and efficient biocontrol of plant pathogenic fungi by Bacillus strains, as evidence of lower disease incidence and severity. For that reason, it is suggested that B. subtilis can be incorporated to integrated management disease, where these strain may be used as biocontrol agent as well as biofertilizer.

Acknowledgements

F.C.R., wants to thank to CONACYT for all financial support during his postgraduate studies.

Author details

Hernández F.D. Castillo[1], Castillo F. Reyes[2], Gallegos G. Morales[1],
Rodríguez R. Herrera[3] and C. Aguilar[3]

1 Universidad Autónoma Agraria Antonio Narro, México

2 Instituto de Investigaciones Forestales Agrícolas y Pecuarias, México

3 Universidad Autónoma de Coahuila, México

References

[1] Velásquez, V.R., Medina A.M.M., Luna R.J.J. (2001). Sintomatología y géneros de pa-
     tógenos asociados con las pudriciones de la raíz del chile (*Capsicum annuum* L.) en el
     norte centro de México. Revista Mexicana de Fitopatología, 19:175-181.

[2] Sosa A., Baro Y. Gonzales M. 2008. Aislamiento, identificación y caracterización de
     cepas de *Bacillus* spp. Con potencialidades para el control biológico de los géneros
     Rhizoctonia, Sclerotium y *Pythium*. Taller Latinoamericano de Biocontrol de Fitopató-
     genos, 2, Ciudad de La Habana, Cuba, 22-29 v. 14(1) p. 63.

[3] Ristaino, J. B. (1991). Influence of rainfall, drip irrigation, and inoculum density on
     the development of *Phytophthora* root and crown rot epidemics and yield in bell pep-
     per. Phytopathology, 81(8):922-929.

[4] Parra, G, Ristaino, J. (2001). Resistance to Mefenoxam and Metalaxyl among field iso-
     lates of *Phytophthora capsici* causing *Phytophthora* Blight of bell pepper. Plant Disease
     85(10):1069-1075.

[5] Hernández-Castillo F.D., Carvajal C.R., Guerrero E., Sánchez A., Gallegos G., Lira-
     Saldivar R.H. (2005). Susceptibilidad a fungicidas de grupos de anastomosis del hon-
     go *Rhizoctonia solani* Khün colectados en zonas paperas de Chihuahua, México.
     International Journal of Experimental Botany, 74(1):259-269.

[6] Schisler, D. A., Slininger, P. J., Behle, R. W., Jackson, M. A. (2004). Formulation of
     Bacillus spp. for biological control of plant diseases. Phytopathology, 94(11):1267-
     1271.

[7] Sid A.A., Ezziyyani, M., Pérez-Sanchez, C., Candela, M.E., (2003). Effect of chitin on
     biological control activity of *Bacillus* spp. and *Trichoderma harzianum* against root rot
     disease in pepper (*Capsicum annuum*) plants. European Journal of Plant Pathology,
     109(6):633-637.

[8] Guédez, C., Castillo, C., Cañizales, L., Olivar, R. (2008). Biological control a tool for
     sustaining and sustainable development. Control Biológico 7(13):50–74.

[9] Cruz, R., Hernández-Castillo, F.D., Gallegos-Morales, G., Rodríguez-Herrera, R.,
     Aguilar-González, C.N., Padrón- Corral, E., Reyes-Valdés, M.H. (2006). *Bacillus* spp.
     como biocontrol en un suelo infestado con *Fusarium* spp., *Rhizoctonia solani* Kühn y
     Phytophthora capsici Leonian y su efecto en el desarrollo y rendimiento del cultivo de
     chile (*Capsicum annuum* L.). Revista Mexicana de Fitopatología, 24(2):105-114.

[10] AL-Janabi, A.A.H.S. (2006). Identification of bacitracin produced by local isolate of
     Bacillus licheniformis. African Journal of Biotechnology, 5(18):1600-1601.

[11] Li, J., Yang, Q., Zhao, L., Zhang, S., Wang, Y., Zhao X., (2009). Purification and char-
     acterization of a novel antifungal protein from *Bacillus subtilis* strain B29. J Zhejiang
     Univ Sci B. 10(4):264–272.

[12] Doornbos, R.F., Loon, L. C., Bakker, P. A. H. M. (2012). Impact of root exudates and plant defense signaling on bacterial communities in the rhizosphere. Agron. Sustain. Dev. 32:227–243.

[13] Suarez-López F., (2010). Evaluación de microorganismos promotores de crecimiento en jitomate (L. *esculentum* L.) bajo condiciones de invernadero. Tesis de Licenciatura, Universidad Autónoma Agraria Antonio Narro, Buenavista, Saltillo, Coahuila.

[14] Kloepper, J. W., Ryu, C.-M., Zhang, S. (2004). Induced systemic resistance and promotion of plant growth by *Bacillus* spp. Phytopathology, 94(11):1259-1266.

[15] Fritze, D. (2004). Taxonomy of the genus *Bacillus* and related genera: the aerobic endospore-forming bacteria. Phytopathology, 94(11):1245-1248.

[16] Rashid, M. H., Mori M., Sekiguchi J. (1995). Glucosam inidase of *Bacillus subtilis*: cloning, regulation, primary structure and biochemical characterization. Microbiology, 141(10):2391-2404.

[17] Beneduzi, A., Passaglia, L.M.P. (2011). Genetic and phenotypic diversity of plant growth promoting Bacilli, in Bacteria in Agrobiology: Plant Growth Responses, Dinesh K. Maheshwari (Ed), Springer-Verlag Berlin Heidelberg. USA. Pp.1-14

[18] Longan N.A., Halket, G. (2011). Developments in the taxonomy of aerobic, endospores-forming bacteria, in: Endospore-forming soil Bacteria. Longan and De Vos (Eds). Espirnger-Verlag, German.

[19] Collins, D. P., Jacobsen, B. J. (2003). Optimizing a *Bacillus subtilis* isolate for biocontrol of sugar beet *Cercospora* leaf spot. Biol. Control, 26(2):153-161.

[20] Zhang, J. X., Xue, A. G., Tambong, J. T. (2009). Evaluation of seed and soil treatments with novel *Bacillus subtilis* strains for control of soybean root rot caused by *Fusarium oxysporum* and *F. graminearum*. Plant Dis. 93(12):1317-1323.

[21] Driks, A. 2004. The *Bacillus* spore coat. Phytopathology, 94(11):1249-1251.

[22] Martin, P. A., Travers, R. S. (1989). Worldwide abundance and distribution of *Bacillus thuringiensis* isolates. Appl Environ Microbiol., 55(10):2437-2442.

[23] [23] Bernhard, K., Jarrett, P., Meadows, M., Butt, J., Ellis, D. J., Roberts, G. M., Pauli, S., Rodgers, P., Burges, H. D. (1997). Natural isolates of *Bacillus thuringiensis*: worldwide distribution, characterization, and activity against insect pests. J Invertebr Pathol., 70(1):59-68.

[24] Armengol, G., Escobar, M. C., Maldonado, M. E., Orduz, S. (2007). Diversity of Colombian strains of *Bacillus thuringiensis* with insecticidal activity against dipteran and lepidopteran insects. J Appl Microbiol., 102(1):77-88.

[25] Yara, K., Kunimi, Y., Iwahana, H. (1997). Comparative studies of growth characteristic and competitive ability in *Bacillus thuringiensis* and *Bacillus cereus* in soil. Appl Entomol Zool., 32(4):625-634.

[26] Borrego, S.F., Perdomo, I., J. de la Paz, Gómez de Saravia, S.G., Guiamet, P.S. (2011). Relevamiento microbiológico del aire y de materiales almacenados en el Archivo Histórico del Museo de La Plata, Argentina y en el Archivo Nacional de la República de Cuba. Revista del Museo de La Plata, 18(19):1-18.

[27] Porcar, M., Caballero, P. (2000). Molecular and insecticidal characterization of a *Bacillus thuringiensis* strain isolated during a natural epizootic. J Appl Microbiol., 89(2): 309-316.

[28] Meadows, M. P., Ellis, D. J., Butt, J., Jarrett, P., Burges, H. D. (1992). Distribution, frequency, and diversity of *Bacillus thuringiensis* in an animal feed mill. Appl Environ Microbiol., 58(4):1344-1350.

[29] Kaelin, P., Gadani, F. (2000). Occurrence of *Bacillus thuringiensis* on cured tobacco leaves. Curr Microbiol., 40(3):205-209.

[30] Damgaard, P. H., Larsen, H. D., Hansen, B. M., Bresciani, J., Jorgensen, K. (1996). Enterotoxin-producing strains of *Bacillus thuringiensis* isolated from food. Lett Appl Microbiol., 23(3):146-150.

[31] Akhurst, R. J., Lyness, E. W., Zhang, Q. Y., Cooper, D. J., Pinnock, D. E. (1997). A 16S rRNA gene oligonucleotide probe for identification of *Bacillus thuringiensis* isolates from sheep fleece. J Invertebr Pathol., 69(1):24-30.

[32] Maeda, M., Mizuki, E., Nakamura, Y., Hatano, T., Ohba, M. (2000). Recovery of *Bacillus thuringiensis* from marine sediments of Japan. Curr Microbiol., 40(6):418-422.

[33] Damgaard, P. H., Granum, P. E., Bresciani, J., Torregrossa, M. V., Eilenberg, J., Valentino, L. (1997a). Characterization of *Bacillus thuringiensis* isolated from infections in burn wounds. FEMS Immunol Med Microbiol., 18(1):47-53.

[34] Damgaard, P. H., Hansen, B. M., Pedersen, J. C., Eilenberg, J. (1997b). Natural occurrence of *Bacillus thuringiensis* on cabbage foliage and in insects associated with cabbage crops. J Appl Microbiol., 82(2):253-258.

[35] Rizali, A., Asano, S., Sahara, K., Bando, H., Lay, B. W., Hastowo, S., Iizuka, T. (1998). Novel *Bacillus thuringiensis* serovar *aizawai* strains isolated from mulberry leaves in Indonesia. Appl Entomol Zool., 33(1):111-114.

[36] Kaur, S., Singh, A. (2000). Natural occurrence of *Bacillus thuringiensis* in leguminous phylloplanes in the New Delhi region of India. World J Microbiol Biotechnol., 16(7): 679-682.

[37] Nair, J. R., Singh, G., Sekar, V. (2002). Isolation and characterization of a novel *Bacillus* strain from coffee phyllosphere showing antifungal activity. J Appl Microbiol., 93(5):772-780.

[38] Collier, F. A., Elliot, S. L., Ellis, R. J. (2005). Spatial variation in *Bacillus thuringiensis/cereus* populations within the phyllosphere of broad-leaved dock (*Rumex obtusifolius*) and surrounding habitats. FEMS Microbiol Ecol., 54(3):417-425.

[39] Priest, F.G. 1989. Isolation and identification of aerobic endospore-forming bacteria, in; Bacillus. Harwood, C.R. (Ed). Plenum press, USA. 27-56 pp.

[40] Basurto-Cadena M. G. L., Vázquez-Arista M., García-Jiménez J., Salcedo-Hernández R., Bideshi D. K, Barboza-Corona J. E. (2012). Isolation of a new Mexican strain of *Bacillus subtilis* with antifungal and antibacterial activities. The Scientific World Journal, Vol. 2012. 7 pages Doi:10.1100/2012/384978.

[41] Carrillo C, Teruel JA, Aranda FJ and Ortiz A. (2003). Molecular mechanism of membrane permeabilization by the peptide antibiotic surfactin. Biochim. Biophys. Acta., 16(11):91- 97.

[42] Nalisha, I., Muskhazli, M., Nor Farizan, T. (2006). Production of Bioactive Compounds by *Bacillus subtilis* against *Sclerotium rolfsii*. Malaysian Journal of Microbiology, 2(2):19-23.

[43] Haleem Khan, A. A, Naseem, Rupa, L., Prathibha, B. (2011). Screening and potency evaluation of antifungal from soil isolates of *Bacillus subtilis* on selected fungi. Advanced Biotech., 10(7):35-37.

[44] Hernández, C.F.D., Aguirre, A.A., Lira, S.R.H., Guerrero, R.E., Gallegos, M.G. (2006). Biological efficiency of organic biological and chemical products against *Alternaria dauci* Kühn and its effects on carrot crop. International Journal of Experimental Botany, 75:91-101.

[45] Bottone, E.J., Peluso, R.W. (2003). Production by *Bacillus pumilus* (MSH) of an antifungal compound that is active against *Mucoraceae* and *Aspergillus* species: preliminary report. J Med Microbiol., 52(1):69-74.

[46] Hernández Castillo F. D., Lira Saldivar, R.H., Cruz Chávez L., Gallegos Morales G., Galindo Cepeda M.E., Padrón Corral E., Hernández Suárez M. 2008. Antifungal potential of *Bacillus* spp. strains and *Larrea tridentata* extract against *Rhizoctonia solani* on potato (*Solanum tuberosum* L.) crop. International Journal of experimental 77(1): 241-252.

[47] Raupach, G. S., Kloepper, J. W. (1998). Mixtures of plant growth-promoting Rhizobacteria enhance biological control of multiple cucumber pathogens. Phytopathology, 88(11):1158-1164.

[48] Martin F. N., Bull C. T. (2002). Biological approaches for control of root pathogens of strawberry. Phytopathology, 92(12):1356-1362.

[49] McSpadden G. B. B., Driks A. (2004). Overview of the Nature and Application of Biocontrol Microbes: *Bacillus* spp. Phytopathology, 94(11):1244-1244.

[50] Klich, M.A., Arthur, K.S., Lax A.R., Blang, J.M. (1994). Iturin A: a potential new fungicide for soared grains. Mycopathologia, 127:123-127.

[51] Katz E., Demain A.L. (1977). The peptide antibiotics of *Bacillus*: chemistry, biogenesis, and possible functions. Bacteriol. Rev., 41:449-474.

[52] Romero, D., de Vicente A., Rakotoaly, R. H., Dufour, S. E., Veening, J-W, Arrebola, E., Cazorla, F.M., Kuipers, O.P., Paquot, M., Pérez-García, A. (2007). The iturin and fengycin families of lipopeptides are key factors in antagonism of *Bacillus subtilis* toward *Podosphaera fusca*. Molecular Plant-Microbe Interactions, 20(4):430-440.

[53] Zeriouh, H., Romero, D., García-Gutiérrez, L., Cazorla, F.M., de Vicente, A. and Pérez-García, A. (2011). The iturin-like lipopeptides are essential components in the biological control arsenal of *Bacillus subtilis* against bacterial diseases of cucurbits. Molecular Plant-Microbe Interactions, 24(12):1540-1552.

[54] Leclère V., Béchet, M., Adam A., Guez, J.S., Wathelet, B., Ongena, M., Thonart, P., Gancel, F., Chollet-Imbert, M. and Jacques, P. (2005). Mycosubtilin Overproduction by *Bacillus subtilis* BBG100 Enhances the Organism's Antagonistic and Biocontrol Activities. Appl. Environ. Microbiol. 71(8):4577-4584

[55] Besson F, Michel G. (1990). Mycosubtilins B and C: minor antibiotics from mycosubtilin-producer *Bacillus subtilis*. Microbios. 62(251):93-9.

[56] Moyne, A.-L., Shelby, R., Cleveland, T.E., Tuzun, S. (2001). Bacillomycin D: an iturin with antifungal activity against *Aspergillus flavus*. Journal of Applied Microbiology, 90:622-629.

[57] Besson F.F., Peypoux G., Michel G., Delcambe L. (1977). The structure of Bacillomycin L, an antibiotic from *Bacillus subtilis*. Eur. J. Biochem., 77:61-67.

[58] Kluge, B., Vater, J., Salnikow, J., Eckart, K. (1988). Studies on the biosynthesis of surfactin, a lipopeptide antibiotic from *Bacillus subtilis* ATCC 21332. FEBS Letters, Vol. 231(1):107-110.

[59] Al-Ajlani, M.M, Sheikh, M.A, Ahmad, Z., Hasnain, S. (2007). Production of surfactin from *Bacillus subtilis* MZ-7 grown on pharmamedia commercial medium. *Microbial Cell Factories, Vol. 6, No. 17, (Jun, 2007)*, pp. (doi: 10.1186/1475-2859-6-17), ISSN 1475-2859

[60] Rezuanul Islam, Md., Yong Tae Jeong, Yong Se Lee, Chi Hyun Song. (2012). Isolation and identification of antifungal compounds from *Bacillus subtilis* C9 inhibiting the growth of plant pathogenic fungi. Mycobiology, 40(1):59–66.

[61] Hassi M, David S, Haggoud A, El Guendouzi S, El Ouarti A, Souda SI, Iraqui M. (2012).*In vitro* and intracellular antimycobacterial activity of a *Bacillus pumilus* strain. African Journal of Microbiology Research 6(10):2299-2304.

[62] Sadfl, N., Chérif, M., Hajlaoui, M.R., Boudabbous, A., Bélanger R. (2002). Isolation and partial purification of antifungal metabolites produced by *Bacillus cereus*. Ann. Microbiol., 52:323-337.

[63] Schallmey, M., Singh, A, Owen P. W. (2004) Developments in the use of *Bacillus* species for industrial production. Can. J. Microbiol. 50: 1–17.

[64] Kugler, M., Loef er, W., Rapp, C., Kern, A., Jung, G. (1990). Rhizocticin A, an antifungal phosphono-oligopeptide of Bacillus subtilis ATCC 6633: biological properties. Archives of Microbiology, 153(3):276-281.

[65] Priest, F.G. (1977). Extracellular enzyme synthesis in the genus Bacillus. Bacteriol. Rev. 41(3):711-753.

[66] Pelletier A., Sygusch J., 1990. Purification and characterization of three chitosanase activities from Bacillus megaterium P1. Applied and Environmental Microbiology, 56(4):844-848.

[67] Pleban, S., Chernin L., Shet I., 1997. Chitinolytic activity of an endophytic strain of Bacillus cereus. Letters in Applied Microbiology, 25(1):284-288.

[68] Sadfi, N., Chérif, M., Fliss, I., Boudabbous, A., H. Antoun, (2001). Evaluation of Bacillus isolates from salty soils and Bacillus thuringiensis strains for the biocontrol of Fusarium dry rot of potato tubers. Journal of Plant Pathology, 83:101–118.

[69] Fiddman, P.J., Rossall, S. (1993). The production of antifungal volatiles by Bacillus subtilis. J. Appl. Bacteriol. 74:119-126.

[70] Essghaier, B., Hedi, A., Hajlaoui, M.R., Boudabous A., Sadfi-Zouaoul, N. (2012). In vivo and in vitro evaluation of antifungal activities from a halotolerant Bacillus subtilis strain J9. African Journal of Microbiology Research 6(19):4073-4083.

[71] Kumar, A., Prakash, A., Johri, B.N. (2011).Bacillus as PGPR in crop ecosystem, in: Bacteria in agrobiology: Crop ecosystem. D.K. Maheshwari (ed). Springer-Verlang Berlin 37-59 pp.

[72] Zhang Y, Jiang L, Ren F, Yang J., Guo H. (2012). Response surface methodology analysis to improve production of tyrothricin in Bacillus brevis. African Journal of Biotechnology 11(47):10744-10752.

[73] Mandic-Mulec, I., Prosser, J. I. 2011. Diversity of endospore-forming bacteria in soil: characterization and driving mechanisms, in endospore-forming soil bacteria, Longan and De Vos (Eds), German, pp. 31-59.

[74] Emberger, O. 1970. Cultivation methods for the detection of aerobic spore-forming bacteria. Zentralbl Bakteriol Parasitenkd Infektionskr Hyg 125 (6):555–565.

[75] Barbeau, B., Boulos, L., Desjardins, R., Coallier, J., Prévost, M.,Duchesne D. (1997). A modified method for the enumeration of aerobic spore-forming bacteria. Canadian Journal of Microbiology, 43(10):976-980.

[76] Jara, S., Madueli, P., Orduz, S., 2006. Diversity of Bacillus thuringiensis strains in the maize and bean phylloplane and their respective soils in Colombia. J App Microbiol., 101(1):117-124.

[77] Berge, O., Mavingui, P., Heulin, T. 2011. Heat treatment (exploring diversity of cultivable aerobic endospore-forming bacteria: from pasteurization to procedures with-

out heat shock selection, in: Endospore-forming Soil Bacteria, Longan and De Vos (Eds), German, pp. 89-113.

[78] Koransky, J.R., Allen, S.D., Dowell, V.R. (1978). Use of ethanol for selective isolation of spore forming microorganisms. Appl. Environ. Microbiol., 35(4):762–765.

[79] Patel, A.K., Ahire, J.J., Pawar, S.P., Chaudhari, B.L., Chincholkar, S.B. 2009. Comparative accounts of probiotic characteristics of *Bacillus* spp. isolated from food wastes. Food Res Int., 42(4):505–510.

[80] Eman, A.H.M, Mikiko, A., Ghanem, K.M., Abdel-Fattah, Y.R., Nakagawa, Y., El-Helow, E.R. 2006. Diversity of *Bacillus* genotypes in soil samples from El-Omayed biosphere reserve in Egypt. J Cult Collect. 5(1):78–84.

[81] Travers, R.S., Martin, P.A., Reichelderfer, C.F. 1987. Selective process for efficient isolation of soil *Bacillus* spp. Appl Environ Microbiol., 53(6):1263–1266.

[82] Parvathi A., Krishna, K., Jose, J., Joseph, N., Nair S. (2009). Biochemical and molecular characterization of *Bacillus pumilus* isolated from coastal environment in Cochin, India. Braz. J. Microbiol., 40(2):269-275.

[83] Dworzanski, J.P., Berwald, L., Meuzelaar, H.L. C. (1990). Pyrolytic methylation-gas chromatography of whole bacterial cells for rapid profiling of cellular fatty acids. Appl. Environ. Microbiol. 56(6):1717-1724.

[84] Logan, N.A., Forsyth, G., Lebbe, L., Goris, J., Heyndrickx, M., Balcaen, A.,Verhelst, A., Falsen, E., Ljungh, A., Hansson, H.B., De Vos, P. (2002). Polyphasic identification of *Bacillus* and *Brevibacillus* strains from clinical, dairy and industrial specimens and proposal of *Brevibacillus invocatus* sp. nov.. Int J Syst Evol Microbiol., 52(3):953-966.

[85] Woese CR. 1987. Bacterial Evolution. Microbiol Rev, 51(2):221-271.

[86] Ash, C.; Farrow, J.A.; Dorsch, M.; Stackebrandt, E.; Collins, M.D. (1991). Comparative analysis of *Bacillus anthracis, Bacillus cereus,* and related species on the basis of reverse transcriptase sequencing of 16S rRNA. Int. J. Syst. Bacteriol., 41(3):343-346.

[87] Wattiau, P., Renard, M.E., Ledent, P., Debois, V., Blackman, G., Agathos, S. N. (2001). A PCR test to identify *Bacillus subtilis* and closely related species and its application to the monitoring of wastewater biotreatment. Applied Microbiology and Biotechnology, 56(5-6):816-819.

[88] Wang, L.T., Lee, F.L., Tai, C.J., Kasai, H. (2007). Comparison of *gyrB* gene sequences, 16S rRNA gene sequences and DNA–DNA hybridization in the *Bacillus subtilis* group. Int J Syst Evol Microbiol, 57(8):1846-1850.

[89] Lihua Lia, Jincai Mab, Yan Lia, Zhiyu Wangc, Tantan Gaoa, Qi Wanga. (2012). Screening and partial characterization of *Bacillus* with potential applications in biocontrol of cucumber *Fusarium* wilt. Crop Protection, 35(1):29–35.

[90] Yamamoto, S.; Harayama, S. (1995). PCR amplification and direct sequencing of gyrB genes with universal primers and their application to the detection and taxonomic analysis of *Pseudomonas putida* strains. Appl. Environ. Microbiol., 61(3):1104-1109.

[91] Castillo, F., Hernández, D., Gallegos, G., Méndez, M., Rodríguez, R., Reyes A.,. Aguilar, C.N. (2010). *In vitro* antifungal activity of plant extracts obtained with alternative organic solvents against *Rhizoctonia solani* Kühn, Industrial Crops and Products, 32(3):324–328.

[92] Van Veen, J.A., Oberbeek, L.S., and Elsas, J.D. 1997. Fate and activity of microorganisms introduced into soil. Microbiology and Molecular Biology Reviews, 61(2): 121-135.

[93] [93] Nava-Díaz, C., Kleinhenz, M.D., Doohan, D.J., Lewis, M.L., Miller, S.A. 1994. *Bacillus* spp. with potential as biological control agents. Phytopathology, 94:74-80.

[94] Wulff, E.G., Mguni, C.M., Mansfeld-Giese, K., Fels, J., Lübeck, M., Hockenhull, J. 2002. Biochemical and molecular characterization of *Bacillus amyloliquefaciens*, *B. subtilis* and *B. pumilus* isolates with distinct antagonistic potential against *Xanthomonas campestris* pv. *campestris*. Plant Pathology, 51(5):574-584.

[95] Yoshida, S., Hiradate, S., Tsukamoto, T., Hatakeda, K., Shirata, A. (2001). Antimicrobial activity of culture filtrate of *Bacillus amyloliquefaciens* RC-2 isolated from mulberry leaves. Phytopathology, 91(2):181-187.

[96] Chiou, A.L., Wu, W.S. 2001. Isolation, identification and evaluation of bacterial antagonists against *Botrytis elliptica* on Lily. Journal of Phytopathology, 149:319-324.

[97] Mari, M., Guizzardi, M., Pratella, G.C. 1996. Biological control of gray mold in pears by antagonistic bacteria. Biological Control, 7(1):30-37.

[98] Harris, A.R., Adkins, P.G. 1999. Biological control of damping-off disease caused by *Rhizoctonia solani* and *Pythium* spp. Biological Control 15:10-18.

[99] Zehnder, G.W., Yao, C., Murphy, J.F., Sikora, E.R., Kloepper, J.W. (2000). Induction of resistance in tomato against cucumber mosaic cucumovirus by plant growth-promoting rhizobacteria. BioControl, 45(1):127-137.

[100] Lucas-García, J.A., Probanza, A., Ramos, B., Ruíz-Palomino, M., Gutiérrez-Mañero, F.J. 2004. Effect of inoculation of *Bacillus licheniformis* on tomato and pepper. Agronomie, 24(4):169-176.

[101] García-Camargo, J. y Díaz-Partida, A. 1991. *Bacillus subtilis* como antagonista de *Fusarium oyxsporum* f. sp. *niveum* y su eficiencia en el control del marchitamiento de la sandía en invernadero. Memorias del XVIII Congreso Nacional de la Sociedad Mexicana de Fitopatología. Puebla de los Angeles, Puebla, México. Resumen, p. 182.

[102] Virgen-Calleros, G., Vázquez-Vázquez, J.L., Anguiano- Ruvalcaba, G.L., Olalde-Portugal, V., Hernández- Delgadillo, R. 1997. Aislamiento de bacterias de la rizósfera de

Capsicum annuum L. antagónicas al desarrollo de *Phytophthora capsici* Leo. Revista Mexicana de Fitopatología, 15:43-47.

[103] Yuen, G.Y., Schroth, M.N., McCain, A.H. (1985). Reduction of *Fusarium* wilt of carnation with suppressive and antagonistic bacteria. Plant Disease, 69(12):1071-1075.

[104] Hernández, S. M., Hernández-Castillo, F.D, Gallegos-Morales, G., Lira-Saldivar, R.H., Rodríguez-Herrera, R., Cristóbal N. Aguilar. 2011. Biocontrol of soil fungi in tomato with microencapsulates containing *Bacillus subtilis*. American Journal of Agricultural and Biological Sciences, 6(2):89-195.

[105] Suárez, A. F. (2011). Manejo de productos biológicos y químicos para el control de *Fusarium oxysporum* y su efecto en el cultivo de melón (*Cucumis melo* L.) en Coahuila, México. Tesis de licenciatura UAAAN, saltillo, Coah, Mexico, pag 55.

Chapter 5

Alternative Weed Control Methods: A Review

G.R. Mohammadi

Additional information is available at the end of the chapter

1. Introduction

Weed interference is one of the most important limiting factors which decrease crop yields and consequently global food production. Weed represent about 0.1% of the world flora and in agroecosystems, weeds and crops have co-evolved together right from the prehistoric times as revealed by pollen analysis studies (Cousens and Mortimer 1995). Weed can suppress crop yield by competing for environmental resources like water, light and nutrients and production of allelopathic compounds. Therefore, weed manage‐ment have been a major challenge for crop producers from the start of agriculture. At the earlier times, since no synthetic chemicals were known, weed control was achieved by some methods such as hand weeding, crop rotation, polyculture and other manage‐ment practices that were low input but sustainable. With the discovery of synthetic her‐bicides in the early 1930s, there was a shift in control methods toward high input and target-oriented ones (Singh et al. 2003).

However, herbicide-reliant weed control methods can cause high costs for crop produc‐ers due to the consumption of fossil fuels (the non-renewable energy resources) (Lybeck‐er et al. 1988). Moreover, ground and surface water pollution by these synthetic chemicals are causes for concern (Hallberg 1989). Fast-developing herbicide-resistant eco‐types of weeds due to increased herbicide application is another serious threat for agri‐culture production (Holt and LeBaron 1990). Therefore, there is an urgent need to develop alternative weed control methods for use in agroecosystems. Many studies have revealed that the alternative methods such as the use of allelopathy phenomenon, cover crops and living mulches, competitive crop cultivars, suitable nutrient management, etc. can be proposed as the low cost, effective and eco-friendly practices for sustainable weed management in cropping systems. In this chapter, the most important alternative weed control methods are discussed.

2. Allelopathy

The term allelopathy was first introduced by Hans Molisch in 1937 and refers to chemical interactions among plants, including those mediated by microorganisms. Rice (1984) defined allelopathy as the effects of one plant (including microorganisms) on another plant through the release of a chemical compounds into the environment. Allelopathy can play a beneficial role in various cropping systems such as mixed cropping, multiple cropping, cover cropping, crop rotations, and minimum and no–tillage systems. The exploitation of allelopathy in agricultural practices as a tool for weed control has shown weed reduction, pathogen prevention and soil enrichment (Kohli et al., 1998).

2.1. Ways by which allelopathy can be used to control weeds in cropping sysyems

In general, the use of allelopathy as a tool to control weeds can be achieved in different ways:

1. Use of crop cultivars with allelopathic properties.

2. Application of residues and straw of allelopathic crops as mulches or incorporated into the soil.

3. Use of an allelopathic crop in a rotational sequence.

4. Application of allelochemicals or modified allelochemicals as herbicides (Kruse et al. 2000).

5. Modification of crops to enhance their allelopathic effects.

2.1.1. Use of crop cultivars with allelopathic properties

It's clear that the crop cultivars differ in their allelopathic ability and thus superior cultivars can be selected for weed management programs (Wu et al. 1999; Olofsdotter et al. 2002). Differences in allelopathic potential between genotypes has been investigated among accessions (genetical different lines or strains of a species) of barley, cucumber (*Cucumis sativus*), oats, soybean (*Glycine max*), sunflower, sorghum (*Sorghum bicolor*), rice and wheat (Copaja et al. 1999, Dilday et al. 1994, Narwal 1996, Miller 1996,Yoshida et al. 1993, Wu et al. 1998).

In a study on 3000 accessions of *Avena* spp. Fay and Duke (1977) found that four accessions apparently exuded up to three times as much scopoletin (a chemical identified as phytotoxic) as a standard oat cultivar. When one of the accessions were grown in sand culture with wild mustard (*Brassica kaber*), the growth of the mustard was significantly less than when it was grown with an accession that exuded a lower amount of scopoletin. In a field experiment, 1000 accessions of rice were screened for allelopathic activity against the two weedy species, barnyardgrass (*Echinochloa crus-galli* Beauv) and *Cyperus difformis*. Of these, 45 accessions showed allelopathic activity against one of the weeds and five accessions inhibited both species (Olofsdotter et al. 1997). Dilday et al. (2001) evaluated 12,000 rice accessions from 110 countries for allelopathy to ducksalad [*Heteranthera limosa* (S.w.) Willd.] and about

5000 have been assessed for allelopathy to redstem (*Ammannia coccinea*) and barnyardgrass. Results indicated that among them, 145 accessions were allelopathic to ducksalad and redstem and 94 accessions demonstrated apparent allelopathic activity to barnyardgrass.

Many weed species are most susceptible to allelochemicals in the seed and seedling stages. Therefore, the ideal allelopathic cultivar must therefore release allelochemicals in bioactive concentrations before the target weeds grow to old. Knowledge about both the critical developmental stage where the crop starts releasing allelochemicals and the critical sensitive stage of the target weeds is therefore essential (Inderjit and Olofsdotter 1998).

2.1.2. Application of residues and straw of allelopathic crops

Weed suppressive effects of crop residues have been explained by different mechanisms, including initial low nitrogen availability following cover crop incorporation (Dyck and Liebman 1994; Kumar et al. 2008; Samson 1991), mulch effects (Mohler 1996; Mohler and Callaway 1991; Mohler and Teasdale 1993), stimulation of pathogens or predators of weed seeds (Carmona and Landis 1999; Conklin et al. 2002; Davis and Liebman 2003; Gallandt et al. 2005; Kremer 1993), and allelopathy (Chou 1999; Weston 1996).

Allelopathic compounds released from crop residues during decomposition can reduce both emergence and growth of weeds. Allelochemicals can be released either through leaching, decomposition of residues, volatilization or root exudation (Chou 1999). In production systems with no-till or conservation tillage that leave nearly all crop residues on the soil surface, the release of allelochemicals from both the growing plants and during residue decomposition could be advantageous (Kruse et al. 2000).

Barnes and Putnam (1983) reported that rye residue used as mulch reduced total weed biomass by 63%. It was found that disappearance of rye allelochemicals was more closely related to weed suppression than to the disappearance of rye residues. Especially due to the massive production of biomass, rye has the potential to influence the growth of succeeding plant species through the release of allelochemicals from the residue (Barnes et al. 1985).

Wheat residues suppress weeds due to the physical effect and to the production of allelochemicals (Petho 1992). their allelopathic effects was positively correlated with the total phenolic content in the tissue of the wheat cultivars (Wu et al 1998). Hydroxamic acids have also been identified in shoot and root tissue of wheat.

The residues of barley have also been associated with phytotoxicity (Overland 1966, Lovett and Hoult 1995). Phytotoxic phenolic compounds, including ferulic, vanillic and phydroxybenzoic acids, have been identified in barley (Börner 1960). The two alkaloids, gramine and hordenine have been confirmed to play an important role in the phytotoxic ability of barley (Lovett and Hoult 1995, Overland 1966). In a study, allelopathic compounds released from residues of barley apparently inhibit the emergence of yellow foxtail (*Setaria glauca*) (Creamer et al. 1996).

In another study, the use of sorghum plant tissues as a mulch or incorporated into the soil led to the reduced weed growth in corn field (Mohammadi et al. 2009). This can be attributed to the allelopathic compounds released from the sorghum plant tissues.

2.1.3. Use of an allelopathic crop in a rotational sequence

The entrance of allelopathic crops into the crop rotations can effectively control weeds. In a study, under reduced or no-till condition a considerable reduction in the population of giant foxtail (*Setaria faberii* Herrm.) was occurred when allelopathic soybean-corn-wheat rotation was followed than in corn alone (Schreiber 1992). In a 5–year field study with sunflower (*Helianthus annuus* L.)–oat rotation, the weed density increase was significantly less in sunflower plots than in control plots (Leather, 1983 a, b; 1987). It was found that sunflower plants possess chemicals, which inhibit the growth of common weed species. Macias et al., (1999) reported some sesquiterpene lactones with germacranolide and guaianolide skeletons and heliannuol from different cultivars of sunflower.

In another study, the inclusion of alfalfa in the crop rotation sequence significantly decreased the interference of weeds in the next crops (Entz et al. 1995). Ominski et al. (1999) conducted a survey in 117 fields in Manitoba, Canada, and found that rotation with alfalfa can effectively reduce the interference of wild oat (*Avena fatua* L.), Canada thistle (*Cirsium arvense* L.), wild mustard (*Brassica kaber* L.) and catchweed bedstraw (*Galium aparine* L.) in the succeeding cereal crops. Therefore, it can be concluded that the inclusion of alfalfa in crop rotation can be an efficient tool in an integrated weed management program. However, climatic and economic conditions are important limiting factors which can notably influence the regional crop rotation scenarios.

2.1.4. Application of allelochemicals or modified allelochemicals as herbicides

A promising way to use allelopathy in weed control is using extracts of allelopathic plants as herbicides (Dayan, 2002; Singh et al., 2005). Because biosynthesized herbicides are easily biodegradable, they are believed to be much safer than synthesized herbicides (Rice, 1984, 1995; Dayan et al., 1999; Duke et al., 2000). Duke et al. (2000) discussed that natural compounds have several benefits over synthetic compounds. For example, natural compounds may have novel structure due to diversity of molecular structure. This diversity is because synthetic chemists have been biased toward certain types of chemistry. They have had almost no interest in water-soluble compounds. Unlike a high proportion of synthetic pesticides, natural compounds are mostly water-soluble and non-halogenated molecules. Natural products relatively have short half-life and therefore considered safe of environmental toxicology standpoint (Duke et al., 2002).

Although, allelochemicals have the potential to be explored as natural herbicides, but prior to using them as herbicides, the following questions should be considered (Bhowmik and Inderjit 2003):

1. At what minimum concentration does each compound have phytotoxic activity?

2. Whether the compound is accurately separated and correctly identified?

3. What is the residence time and fate of the compound in the soil environment?

4. Does the compound influence microbial ecology and physicochemical properties of the soil?

5. What is the mode of action of the compound?

6. Has the compound any adverse effect on desired crops?

7. Whether the compounds are safe from health standpoint?

8. Whether the large production of the compound at commercial scale is economical?

Plant chemicals associated with allelopathic activity have been reported in most cases to be secondary metabolites from shikimic acid, acetate, or terpenoid pathways (Rizvi and Rizvi 1992; Vokou 2007). Some of the natural products exploited as commercial herbicides are triketone, cinmethylin, bialaphos, glufosinate and dicamba. The compounds having potential herbicidal activity but not commercially used are sorgoleone, artemisinin and ailanthone (Bhowmik and Inderjit 2003).

Sorgoleone is an allelochemical of sorghum which constitutes more than 80% of root exudate composition (Nimbal et al., 1996a; Czarnota et al., 2003). This compound inhibited the evolution of O_2 during photosynthesis in potato (*Solanum tuberosum* L.) and in common groundsel (*Senecio vulgaris* L.) (Nimbal et al. 1996a). Nimbal et al. (1996b) carried out a study on sorgoleone using triazine-susceptible potato and redroot pigweed thylakoids. Sorgoleone was a competitive inhibitor of atrazine binding sites. Sorgoleone also inhibited the photosystem II electron transport reactions (Gonzalez et al., 1997).

However, sorghum shoots produce higher amounts of cynogenic glucosides whose phenolic breakdown products inhibit plant growth (Einhellig and Rasmussen, 1989; Weston et al., 1989; Se'ne et al., 2001). In a study, Mohammadi et al. (2009) reported that the spray of sorghum shoot extract (Sorgaab) reduced weed infestation in corn field.

Artemisinin, a sesquiterpenoid lactone is an allelochemical of annual wormwood (*Artemisia annua* L.). It has been shown to inhibit the growth of redroot pigweed, pitted morning glory (*Ipomoea lacunosa* L.), annual wormwood and common purslane (*Purtulaca oleracea* L.) (Duke et al., 1987). Duke et al. (1987) concluded that artemisinin is a selective phytotoxin with herbicidal activity similar to cinmethylin (Bhowmik, 1988).

Ailanthone an allelochemical of *Ailanthus altissima* L. exhibited a strong herbicidal activity when sprayed on soil before the seed germination. It, however, also had dramatic effects when sprayed onto seedlings after their emergence from soil (Bhowmik and Inderjit 2003).

However, most of allelochemicals indicate a poor performance under field conditions compared to laboratory conditions. Moreover, many allelochemicals exhibit rapid dissipation under natural situations and thus fail to give desired results (Singh et al. 2003). Therefore, further studies are needed to enhance performance and stability of allelochemicals under field conditions.

2.1.5. Modification of crops to enhance the allelopathic effect

Breeding of crops for allelopathic ability by using the methods like screening and biotechnology is another promising strategy for efficient weed control. Just as crop plants are bred for disease resistance, crop plants can be bred to be allelopathic to weeds common to specific regions (Rice, 1984, 1995; Jensen et al., 2001; Wu et al., 2000, 2003; Olofsdotter et al., 2002; He et al., 2004). Allelopathic effect against a broad spectrum of weeds has been proposed as a valuable character of an allelopathic crop and the possibility of inserting resistance genes towards one or several weeds as part of a breeding strategy of a crop has been mentioned (Olofsdotter et al. 1997).

Genetic modification of crop plants to improve their allelopathic properties and enhancement of their weed-suppressing ability has been suggested as a possibility (Kruse et al 2000). Use of biotechnological transfer of allelopathic traits between cultivars of the same species or between species has also been proposed (Chou 1999, Macias 1995, Macias et al. 1998, Rice 1984).

Several researchers have suggested improvement of allelopathic properties of crop cultivars by traditional breeding or by genetic manipulation. For example, there has been significant progress in isolating rice allelochemicals (Rimando et al., 2001) and locating genes controlling allelopathic effects of rice (Jensen et al., 2001). These researchers identified quantitative trait loci (QTL) associated with the rice allelochemicals against barnyardgrass. This is an important step toward breeding allelopathic rice varieties. It was found that 35% of the total phenotypic variation of allelopathic activity of population was explained by four main effect QTLs situated on three chromosomes.

In wheat, the control of hydroxamic acid accumulation seems to be multigenic involving several chromosomes. Chromosomes of group 4 and 5B are apparently involved in the accumulation of hydroxamic acids (Niemeyer and Jerez 1997).

In barley, a gramine synthesis gene has been detected on chromosome 5 (Yoshida et al. 1997). Moharramipour et al. (1999) reported that one or two genes control the synthesis of gramine. DIBOA is a hydroxamic acid compound which has been found in wild *Hordeum* species by Barria et al. in 1992 and the production of DIBOA by cultivated barley could possibly be achieved by transferring genetic material from wild barley species (Gianoli and Niemeyer 1998).

Duke et al. (2000) suggested that biotechnology may eventually allow for the production of highly allelopathic crops through the use of transgenes to increase allelochemical production to levels that effectively manage weeds without herbicides or with reduced herbicides input. However, it has been stated by Wu et al. (1999), that even though genetic manipulation seems promising, it might be more feasible to select for crop cultivars with improved allelopathic properties using conventional breeding methods, because of the strict regulation and public concern about transgenic crops.

2.2. Aromatic plants

Aromatic plants could play an important role in the establishment of sustainable agriculture because of their ability to produce essential oils that could be used in the development of biological pesticides (Isman 2000). Eessential oils are increasingly adopted in agriculture for their use as pesticides (Daferera et al. 2003; Isman 2000; Tuncw and Sahinkaya 1998).

Palmer amaranth (*Amaranthus palmeri* L.) germination was inhibited by essential oils of certain aromatic plants including lemon basil (*Ocimum citriodorum* L.), oregano (*Origanum vulgare* L.), and sweet marjoram (*Origanum majorana* L.) (Dudai et al. 1999). Dhima et al. (2009) found that anise, sweet fennel, lacy phacelia, and coriander aqueous extracts inhibited by 100% germination, root length, and seedling fresh weight of barnyardgrass. In another study, Tworkoski (2002) tested 25 plant-derived essential oils for herbicidal activity and found that those from red thyme (*Thymus vulgaris* L.), summer savory (*Satureja hortensis* L.), cinnamon (*Cinnamomum zeylanicum* L.) and clove (*Syzygium aromaticum* L.) were most toxic and caused cell death due to rapid electrolyte leakage on the detached leaves of dandelion (*Taraxacum officinale* L.).

3. Cover crops and living mulches

The term cover crop refers to a plant which is grown in rotation during periods when main crops are not grown. Cover crops are usually killed (mechanically or chemically) before the planting of the main crop. However, living mulches are cover crops that are planted between the rows of a main crop and are maintained as a living ground cover during the growing season of the main crop. Although, living mulches are sometimes referred to as cover crops, they grow at least part of the time simultaneously with the main crop. Apart from the definitions, both cover crop and living mulch suppress weeds by the similar mechanisms.

In general, cover cropping systems have high potentials for weed management in agroecosystems. Cover cropping has long and short-term weed control effects (Barberi 2002) as a result of competition and/or allelopathy exerted by the crop (Randall et al. 1989; Boydston and Hang 1995). These effects can enhance the effectiveness of other non-chemical weed control means in view of an effective integrated approach (Creamer et al. 1996; Bond and Grundy 2001). Long-term weed control effects are due to the prevention of emergence and/or seedling suppression of species of different seasonality compared to the following crop, while short-term effects take place when emergence prevention and seedling suppression occur in species presenting the same seasonality of the following crop (Campiglia et al. 2009).

3.1. How can a cover crop or living mulch affect weed growth?

The effects of a cover crop or living mulch are achieved by a rapid occupation of the open space between the rows of the main crop or generally, the niches that would normally be filled by weeds (Teasdale 1998). This prevents germination of weed seeds and reduces the growth and development of weed seedlings. Generally, the weed suppressing ability of these systems is thought to be based on alleopathic properties, physical impedance of germination and seedling growth, and competition for light, water, and nutrients (Teasdale, 1993; Teasdale and Mohler, 1993).

Germination of weed seeds may be inhibited by complete light interception (Phatak, 1992) by cover crops or by secretion of allelochemicals (White et al., 1989; Overland, 1966). A delay in emergence of weeds because of the presence of living mulches or cover crops can also adversely affect weed seed production. Moreover, the presence of living mulches or cover crops leads to greater seed mortality of weeds by favoring predators (Cromar et al. 1999).

Once established, cover crops and living mulches can use the light, water, and nutritional resources that would otherwise be available to weeds. This can result in the inhibition of weed seed germination and reduction in the growth and development of weed seedlings. Therefore, weeds attempting to establish along with a cover crop or living mulch would be in competition for resources and may not develop sufficiently. Moreover, physical impediment to weed seedlings is another mechanism by which these crops suppress weeds (Facelli and Pickett 1991; Teasdale 1996; Teasdale and Mohler 1993).

Since, most crop-living mulch systems are sufficiently supported by water and nutrients, it seems that light is the most important resource for competition between living mulches and weeds. In a study, Kruidhof et al. (2008) found that weed suppression is positively correlated to early light interception by the living mulch and is sustained by the strong negative correlation between cumulative light interception and weed biomass. Similarly, Steinmaus et al. (2008) reported that weed suppression was linked to light interception by the mulch cover for most weed species.

Allelopathy is another mechanism by which living mulches may suppress weeds (Fujii 1999). However, this is difficult to separate experimentally from mechanisms relating to competition for growth resources. Allelopathic compounds can be released into the soil by a variety of mechanisms that include decomposition of residues, root exudation, and volatilization (Weston 2005). According to Westra (2010) root exudation produces allelopathic compounds that are actively secreted directly into the soil rhizosphere by living root systems. The allelochemicals then move through the soil by diffusion and come into contact neighboring plants. This creates a radius effect, where proximity to the allelopathic species results in greater concentrations of the allelochemical, which, in turn, typically decreases the growth of neighboring plants.

3.2. Factors influencing the weed suppressive ability of a cover cropping system

The success of a cover cropping system to suppress weeds can be influenced by some factors including:

3.2.1. Cover crop species

The variability among cover crop species determines the importance and opportunities of species selection as a component in the design of a suitable weed management system (den

Hollander et al. 2007). Cover crop species are significantly different in their ability to suppress weeds. In a study, among six leguminous species (Persian clover, *Trifolium resupinatum* L.; white clover, *T. repens* L.; berseem clover, *T. alexandrinum* L.; hairy vetch; alfalfa; and black alfalfa, *M. lupulina* L.), the lowest weed dry weights were obtained from the plots interseeded with hairy vetch as compared with the other species (Mohammadi 2009).

Morphological growth characteristics, such as early relative growth rate of leaf area and earliness of height development, have been identified to determine competition in intercropping systems (Kropff and van Laar 1993). Weed suppression benefits from a rapid soil cover, as this reduces the germination and establishment of weeds as well as the relative competitive ability of established weed seedlings (Ross and Lembi, 1985). The relative growth rate reflects the increase of characteristics like soil cover and dry matter accumulation during early development, when growth is still exponential. The relative growth rate of a plant species is thus affected by its light capturing ability, by the efficiency by which it converts light into biomass and by the fraction of newly produced biomass which is invested in leaves (den Hollander et al. 2007).

For weed competition and weed suppression, earliness has also been reported an important characteristic (De Haan et al., 1994). Particularly for competition for light, which is asymmetric (Weiner, 1986), obtaining a good starting position seems highly relevant. From that perspective the relative growth rate seems to be a more important characteristic than the maximum accumulated amount of biomass (den Hollander et al. 2007).

Apart from soil cover development, height is also an important characteristic, determining competition for light (Berkowitz, 1988). Akanvou et al. (2001) found that *Crotalaria juncea*, *Cajanus cajan* and *M. pruriens* can be considered as species with a higher competitive ability than *Calopogonium mucunoides*, *Stylosanthes hamata* and *A. histrix*. This was explained by the combination of high initial growth rates for height and leaf area development. Additionally, the high final height of *C. juncea* and *C. cajan* may confer higher competitiveness throughout the growing season.

However, differences in soil cover development do not only depend on species differences in morphology and physiology. The relative starting position, determined by, for instance, seed size, seeding rate and date of emergence, is another major factor in this respect. As small-seeded species may be more sensitive to conditions that might cause a poor establishment (den Hollander et al. 2007).

In general, ideal cover crops or living mulches for weed suppression should have the following characteristics:

a. Ability to provide a complete ground cover of dense vegetation.

b. Rapid establishment and growth that develops a canopy faster than weeds.

c. Selectivity between suppression of weeds and the associated crop (Teasdale 2003).

3.2.2. Agricultural practices

Among agricultural practices, both time and rate of living mulch or cover crop planting can be important factors determining the success of a these systems to suppress weeds. In a study, Mohammadi (2010) observed that increased hairy vetch planting rate from 0 to 50 kg ha^{-1} improved corn yield (by 11%) and reduced weed dry weight (by 50.9%). In another study, increasing berseem clover (as a living mulch) planting rate from 20 to 40 kg ha^{-1} reduced weed density and biomass by 20.2 and 10.1% respectively, in corn field (Mohammadi et al. 2012). It was hypothesized that as living mulch density is increased, canopy closure would occur more rapidly, decreasing the amount of photosynthetically active radiation (PAR) available beneath the canopy. This would result in a concomitant decrease in weed biomass until an optimum living mulch density is achieved, beyond which, no further decrease in weed biomass could be obtained (Collins et al. 2008).

Planting time of a cover crop or living mulch is also a very important factor which affects its weed suppressive ability. For example, when rye (*Secale* sp.) or small-grain living mulches were interseeded at or near planting of the main crop, they could provide higher levels of weed suppression (Rajalahti et al. 1999; Brainard and Bellinder 2004). Mohammadi et al. (2012) also reported that the lowest weed density and biomass were occurred in the treatment in which berseem clover (as a living mulch) was interseeded 15 days before corn planting as compared with the other interseeding times (simultaneous with or 15 days after corn planting). This may be related to the faster occupation of the open space between the rows of the main crop and consequently a more efficient use of the environmental resources by the living mulch which can ultimately lead to the reduced weed growth and development.

Generally, the biomass produced by a living mulch highly depends on its planting rate and time. Moreover, there is often a negative correlation between living mulch and weed biomasses (Akemo et al. 2000; Ross et al. 2001; Sheaffer et al. 2002). Similar findings were also reported by other researchers (Meschede et al. 2007; Mohammadi 2010; Mohammadi et al. 2012).

However, an important concern on cover cropping systems is reduced main crop yield due to competition or allelopathic effects of living mulches or cover crops. Therefore, appropriate management of these systems is very critical to reduce harmful effects of cover crops or living mulches on main crop while allowing them to grow sufficiently to reap potential benefits. The selection of suitable cover crop and living mulch species is very important. Ideally, the main crop and the cover crop should differ to a high degree in the way they explore resources, thus avoiding competition between both species to at least some extent (Vandermeer, 1989). Generally, greater potential benefits might be expected from living mulches with a very different active growth period than the main crop.

Other attempts have also been made to reduce the unsuitable effects of the cover crops while maintaining their weed suppressing ability. Brandsaeter and Netland (1999) focused on temporal complementarity by separating periods of vigorous growth of the cover crop and the main crop, while Vrabel (1983) used chemical control of the cover crop to reduce yield losses. Brainard et al. (2004) evaluated different options, particularly, cover crop species, time of seeding, use of supplemental nitrogen and herbicide regulation. Ross et al. (2001) conducted mechanical control of the cover crop and combined this with a screening of different cover crops.

It can be concluded that, although, inclusion of cover crops and living mulches in cropping systems can be useful to reduce harmful effects of weeds, but an appropriate management program is very essential to obtain the best results.

(For more information on the weed suppressing role of living mulches and their management in cropping systems, please see Mohammadi (2012), Living mulch as a tool to control weeds in agroecosystems: A Review).

4. Planting arrangement

An integrated weed management approach should employ multiple control strategies (Walker and Buchanan 1982), possibly including the development of weed suppressing cropping systems (Shrestha and Fidelibus 2005). Alteration of planting arrangement can be proposed as an efficient practice to suppress weeds in agroecosystems. This can be achieved by the change of planting density, row spacing, row orientation, etc.

4.1. Planting density and row spacing

The practice of increasing crop plant density by using higher seeding rates associated with narrower row spacing can lead to earlier canopy closure, thus shading weeds in their early developmental stages (Vera et al. 2006). Sharma and Angiras (1996 a,b) and Angiras and Sharma (1996) found that reduced row spacing increased light interception by crops and reduced weed biomass, increasing crop yield. The studies conducted on barley (*Hordeum vulgare* L.) have shown that higher seeding rates using cultivars with differing competitive abilities enhanced crop competitiveness against wild oat (*Avena fatua* L.) (Harker et al. 2009; Watson et al. 2006; O'Donovan et al. 2000).

In general, increasing crop seeding rates can hasten and increase resource use, and thereby reduce the negative effect of weeds (Berkowitz 1988; Mohler 1996). Therefore, weed management and cereal and pulse crop yields were improved with higher than recommended seeding rates in the absence of herbicides (Barton et al. 1992; Kirkland 1993; Townley-Smith and Wright 1994; Khan et al. 1996; Ball et al. 1997; O'Donovan et al. 2000). In another study, tartary buckwheat (*Fagopyrum tataricum*) was effectively suppressed when canola (*B. rapa*) seeding rate was increased from 2 to 8 kg ha^{-1} (O'Donovan and Newman 1996).

Weed suppression by crops appears to be enhanced by size-asymmetric competition, in which the larger crop plants suppress the initially smaller weed plants (Schwinning and Weiner 1998; Weiner 1990). At high-density, size-asymmetric competition is stronger and starts earlier, whereas the crop still has a large size advantage. At relatively low crop densities, crop cover early in growing season is low, leaving a larger amount of resources available for the weeds, thus enabling them to establish and grow quickly (Kristensen et al. 2008).

Row spacing can also affect the crop competitive ability against weeds. In a study, rice grown in 30-cm rows had greater weed biomass and less grain yield than in15-cm and 10–20–10-cm rows and crops in the wider spacing (30-cm) were vulnerable to weed competition for the longest period (Chauhan and Johnson 2011). In another study, Mohammadi et al. (2012) reported that corn yield improved and weed biomass reduced in response to increasing plant density and decreasing row spacing.

Row spacing can also influence the critical period of weed control in crops. It is hypothesized that narrow row spacings may decrease the interval of critical weed competition periods (Chauhan and Johnson 2011). According to Chauhan and Johnson (2011) the critical weed-free periods for rice planted at the 30-cm rows were up to 8 days longer than the other two rows spacings (15-cm and 10–20–10-cm rows).

Moreover, several studies have documented the reduced competitive ability of short-stature cultivars (Harker et al., 2009; O'Donovan et al., 2000) and improvements in the competitive ability of shorter varieties could be derived from narrower row spacing (Drews et al., 2004).

In general, the higher weed densities typical in low-input and organic systems may make narrow row spacing and higher planting density particularly attractive.

4.2. Row orientation

Light is an important determinant of crop productivity. Crops can be manipulated to increase shading of weeds by the crop canopy, to suppress weed growth, and to maximize crop yield (Borger et al. 2010). In general, cropping systems that reduce the quantity and quality of light in the weed canopy zone suppress weed growth and reduce competition (Borger et al. 2010; Crotser and Witt 2000; Rajcan et al. 2002; Sattin et al. 1994; Shrestha and Fidelibus 2005; Teasdale 1995). During early growth stages, there is interference between crop and weed plants because of reflected light. The reflection of far-red photons by the stem of one plant lowers the red to far red photon ratio of light experienced by the stems of neighboring plants. This modifies the light environment in the plant stem tissue, which results in an increased stem elongation rate. As plants age, the crop canopy closes, and mutual shading further increases the competition for photosynthetic light (Borger et al. 2010).

One possible way to reduce light interception by weeds and to increase light interception by the crop canopy is to manipulate the crop row orientation (Holt 1995). So that, orientating crop rows at a near right angle to the sunlight direction increases the shading of weeds between the rows. In a study, within wheat and barley crops oriented east–west, weed biomass was reduced by 51 and 37%, and grain yield increased by 24 and 26% (compared with crops oriented north–south). This reduction in weed biomass and increase in crop yield likely resulted from the increased light (photosynthetically active radiation) interception by crops oriented east–west (i.e., light interception by the crop canopy as opposed to the weed canopy was 28 and 18% greater in wheat and barley crops oriented east–west, compared with north–south crops) (Borger et al. 2010).

According to Alcorta et al. (2011), rows oriented east-west allowed less light penetration to the weed canopy zone than north-south rows throughout the growing season and weed species responded to low light levels by producing leaves with larger specific leaf area and leaf area ratios than those in the north-south rows. Moreover, the leaf, stem and root dry weight of the weed species in the east-west rows was reduced by 30% compared to the weed species in north-south rows.

According to the results of another study, during periods of peak PAR, the Red : Far Red (R : FR) ratio was more than three times greater under the grape canopy in north-south rows than in east-west rows, indicating that row orientation can affect both quantity and quality of light available to weeds (Shrestha and Fidelibus 2005).

However, the effect of row orientation can vary with latitude and with the seasonal tilt of the earth in relation to the sun. Near the equator, north–south (as opposed to east–west) orientation gives crops higher levels of light absorption for most of the year. At higher latitudes (up to 55°), absorption is highest in north–south crops in summer and east–west crops for the rest of the year. From 65° upwards, east–west orientation gives greatest light absorption all year (although the difference between orientations is minor) (Mutsaers 1980).

It can be concluded that manipulation of row orientation can be an ideal method to incorporate into an integrated weed-management program because it does not cost growers anything to implement, and it is environmentally friendly compared with chemical weed control tactics (Mohler 2001). However, the geographical and seasonal conditions should be considered.

4.3. Spatial uniformity

According to Kristensen et al. (2008) increased crop density and spatial uniformity can play an important role in weed management and a strategy based on increased crop density and spatial uniformity can reduce or eliminate herbicide application in conventional cereal production. Crop spatial uniformity decreases competition within the crop population early in the growing season (Olsen and Weiner 2007) and maximizes the total shade cast by the crop by reducing self-shading (Weiner et al. 2001). In a study, In the presence of weeds, the highest yields were obtained with high crop density and high spatial uniformity (Kristensen et al. 2008).

However, the early size advantage of the crop is the theoretical basis for our prediction of positive effects of increased density and spatial uniformity on weed suppression (Weiner et al. 2001). Therefore, it can be concluded that increased crop density and uniformity will not lead to effective weed suppression when weeds have the initial size advantage (e.g., perennial weeds), or are able to catch up in size with the crop before competition becomes intense (Kristensen et al. 2008). Moreover, one might expect the effects of high crop density and spatial uniformity on weeds to be more pronounced at low soil nitrogen levels because weeds grow more slowly at low fertilization levels (Blackshaw et al. 2003).

5. Competitive crop cultivars

In a plant community, competition occurs when the environmental resources are limited. Competition for limited resources is the primary causes of crop loss from weeds. Crop cultivars that better compete with weeds or preempt resources from weeds may benefit an integrated weed management program (Jordan 1993; Lemerle et al. 1996; Lindquist and Kropff 1996).

The competitive ability of a plant has two components, the competitive effect—ability of an individual to suppress other individuals—and the competitive response—ability of an individual to avoid being suppressed—corresponding to different abilities of plants to acquire and use resources (Goldberg, 1990). Competitive effect is related to resource acquisition, with large or tall plants competitively depressing smaller ones (Gaudet and Keddy, 1988; Keddy and Shipley, 1989; Aarssen, 1992). Plants avoid being suppressed by acquiring resources faster (foraging strategy), shifting resource acquisition site or time relative to neighbors (escaping strategy), or conserving scarce resources (persistent strategy) (Navas and Moreau-Richard 2005).

The development of competitive crop cultivars is an important aspect of integrated weed management and can reduce reliance on herbicides (McDonald 2003). The ideal weed competitive cultivars are high-yielding under both weed-free and weedy conditions and have strong weed-suppressive ability. Weed-suppressive ability is the ability to suppress weed growth and reduce weed seed production and, hence, benefit weed management in the subsequent growing season (Jannink et al., 2000; Zhao et al., 2006).

In a general view, crop competitive ability can be divided into two practical perspectives. Crop tolerance is defined as the ability of the crop to endure competitive stress from the presence of weeds without substantial reduction in growth or yield. Weed suppressive ability is the ability of the crop to reduce weed growth and fecundity (So et al. 2009). Weed suppressive ability is determined by assessing weed biomass or weed seeds under weedy conditions. Stronger Weed suppressive ability is not always associated with higher yield under weedy conditions (Saito et al. 2010). However, suppressing weeds reduces weed seed production and benefits weed management in future grow- treating seasons while tolerating weeds only benefits the current growing season. Moreover, weed pressure from unsuppressed weeds increases the likelihood of crop yield loss, irrespective of the crop's tolerance (Jannink et al. 2000).

Ideally, a competitive cultivar should both tolerate weeds and suppress their growth (Jordan 1993). The tolerance of a crop cultivar to weeds is the ability of that cultivar to maintain high seed yields when weeds are present. The weed suppression ability of a crop cultivar is the ability of that cultivar to reduce weed growth and subsequent seed production (Spies et al. 2011). Callaway (1992) documents genetic variability for both perspectives in numerous crop species and many authors suggest breeding to improve the traits (Garrity et al., 1992; Callaway and Forcella, 1993; Kropff and van Laar, 1993; Wortmann, 1993; Liebman and Gallandt, 1997; Bussan et al., 1997).

In general, the traits offering weed competitive ability to crop cultivars can be divided into the several groups including:

5.1. Canopy and morphological traits

Canopy architecture influences many canopy processes including interactions between the crop and specific aspects of its environment (Daughtry et al., 1983; Welles and Norman, 1991). Canopy architecture is a function of leaf number, shape, distribution, orientation, and plant size, which collectively determine the vertical distribution of light within the crop canopy (Williams et al., 1968; Girardin and Tollenaar, 1994).

Sinoquet and Caldwell (1995) reported that light is the primary resource for which weeds will compete in an irrigated and high N-input crop production system. Total canopy leaf area index, height, rate of leaf area development and their distribution in the canopy, are the most important traits in competition for light (Sinoquet and Caldwell, 1995), which can be improved through cultural practices or/and by plant breeding (Lindquist and Mortensen, 1998). Identifications and improvements in traits driving light interception, such as height, leaf area index and canopy diameter can increase competitiveness of several crops (Bennett and Shaw 2000; Callaway 1992; Lindquist and Mortensen 1998; Lindquist et al. 1998). According to Watson et al. (2002) a species competitive strength is strongly determined by its share in leaf area when the canopy closes and interplant competition starts. In general, rapid canopy closure and a large, late-maturing canopy were positively associated with competitive ability (So et al. 2009).

In field pea cultivars, vine length and the leafy characteristic may be important genetic characteristics associated with competition (Spies et al. 2011). However, Wall and Townley-Smith (1996) believed that vine length was more important than the leafed or semi leafless trait. Several traits relate to competitive ability of dent corn, including plant height, shoot growth rate, canopy density (Lindquist and Mortensen 1998), leaf uprightness (Sankula et al. 2004), crop maturity, leaf area growth rate (Begna et al. 2001 a, b), canopy closure, and maximum leaf area index (Lindquist et al. 1998).

In an experiment, when wild-proso millet competed with sweet corn, hybrids with a large canopy were best equipped to tolerate the weed and suppress wild-proso millet growth and seed production, even for late-maturing hybrids that competed the longest period of time (So et al. 2009).

Plant height and tillering ability are also key characteristics for wed suppressive ability under specific growing environments. Their relative contributions to weed suppressive ability could be affected by crop establishment method, agro-ecosystems (upland or lowland) or weed species (Saito et al. 2010). According to Lemerle et al., (1996) greater tiller numbers, taller plants, elevated photosynthetically active radiation interception, and greater early season biomass accumulation were all found in the most competitive genotypes in a study of wheat genotypes from around the world.

Wang et al. (2006) also reported that an erect cowpea genotype is more competitive due to its taller stature, greater height growth rate, and higher position of maximal leaf area densi-

ty, despite a lower photosynthetic rate and light use efficiency than the other cowpea genotypes. They concluded that erect growth habit may be generally more competitive with
weeds compared to semi-erect or prostrate growth habit. In another study, the size of the
flag leaf has been correlated with competitive ability in barley (Watson et al. 2002).

5.2. Phenological traits

Weed species differ markedly in their development phenologies. An effective, 'broad spectrum' weed-suppressive cultivar will therefore need a strong competitive presence over the
full duration of the season. The positive correlation between weed biomass and time to maturity of cultivars in organic fields indicates that weed growth was higher in cultivars with
increased time to maturity. Thus, it may be desirable for organic producers to use early maturing cultivars to reduce weed biomass in the field (Jannink et al. 2001). In other words,
faster time to maturity was found to be associated with reduced weed biomass. Huel and
Hucl (1996) in evaluating 16 genotypes of spring wheat (*Triticum aestivum* L.), found a positive correlation between early maturity and competitive ability. In another study, earlier
flowering cultivars of soybean were more successful at suppressing weed growth.

In general, earlier maturing cultivars might have higher relative growth rate. Across a broad
spectrum of species, researchers have found a negative relationship between leaf longevity
and plant relative growth rate (Reich et al., 1997). They posit that this relationship occurs because long-lived leaves require more structural carbon and protective secondary metabolites
than short-lived leaves, such that the cost to the plant of developing photosynthetic capacity
is greater (Jannink et al. 2001).

Moreover, earlier maturing cultivars might produce larger seeds if they diverted a greater
proportion of photosynthate to their fewer reproductive structures (Kollman et al., 1979;
Openshaw et al., 1979; Wallace et al., 1993). With a larger initial size they might achieve a
higher absolute growth rate despite equal relative growth rate. This can lead to the higher
competitive abilities of these cultivars.

However, this idea contrasts with that of some workers who have suggested that later maturity confers greater competitive ability against weeds because cultivars that remain vegetative grow to be taller (Hinson and Hanson, 1962; McWhorter and Hartwig, 1972; Monks
and Oliver, 1988). Jannink et al. (2000) also found earlier-maturity soybean cultivars displayed greater initial growth and weed suppression, compared to later-maturity cultivars,
but were less able to sustain weed suppression throughout the season due to senescence. It
can be concluded that earlier maturing cultivars have a higher weed suppressive ability if
they can sustain this ability throughout the growing season.

5.3. Growth parameters

In plant ecology, relative growth rate (RGR) is considered to be one of the key characteristics
of plants that is positively correlated with competitive ability (Grime 1977; Grace 1990). Holt
and Ocrutt (1991) showed that the RGR is one of the most important plant growth parameters to increase the competitive ability of cotton against weeds. As a result of a high RGR, a

crop will rapidly increase in size and can occupy a larger space, both below and above ground. Consequently, such a crop has the opportunity to acquire a larger share of limiting resources, such as light, nutrients, and water, than a weed.

In a study conducted by Mohammadi (2007) some plant growth parameters including leaf area index, specific leaf area, crop growth rate, relative growth rate and net assimilation rate were evaluated to identify which of them can enhance corn competitive ability against weeds. The variable selection using the stepwise multiple linear regression method revealed that, among the growth parameters under study, both the relative growth rate and the specific leaf area (SLA) were the best predictors of corn cultivar competitiveness.

Broad surveys across taxa have found SLA to predict relative growth rate (Hunt and Cornelissen, 1997; Reich et al., 1997) and implicate SLA in competitive ability (van der Werf et al., 1993). A high SLA contributed to increased light interception by crops (Jannink et al. 2000) and led to a reduction in the amount of light available to weeds. According to Johnson et al. (1998), a high SLA is one of the best predictors of cultivar competitiveness against weeds. Dingkuhn et al. (1999) also reported that more weed competitive cultivars have a high SLA, which leads to a high leaf area index.

Tollenaar et al. (1994) showed that differences among corn hybrids in competition against weeds can be attributed to differences among them in the leaf area index and the transmission of late-season photosynthetic photon flux density. In the case of interactions with weeds, more rapid early leaf area development, higher leaf area index and biomass accumulation can play important roles. Leaf area and plant height might affect significantly crop–weed interactions (Blackshaw, 1994; Lemerle et al., 1996). Cultivars with strong weed suppressive ability accumulated more biomass, produced more tillers and displayed higher leaf area index during the vegetative growth stage than those with weak weed suppressive ability (Saito et al. 2010).

Plant height is another growth parameter which can influence weed suppressive ability of a crop cultivar. The negative correlation between weed biomass and plant height in organic fields implies that weed biomass decreased as height increased, suggesting that height does help to suppress weeds. In several studies, plant height was associated with competitive ability in both conventional (Huel and Hucl, 1996; Lemerle et al., 1996; Hucl, 1998) and organic systems (Gooding et al., 1993). In soybean, plant height 6 to 7 wk after emergence showed moderately high heritability, strong genetic correlation to weed suppressive ability and was quick and simple to measure. These characteristics make it an ideal indirect selection criterion, particularly in a practical soybean breeding program where labor needs at the time to measure early height are not as high as in the spring or the fall (Jannink et al. 2000).

In rice, under severe weed competition, higher biomass accumulation at 42 days after sowing was associated with higher weedy yield. For adaptation to both moderate and severe weed pressure, cultivars should have high-yielding ability, high plant height at maturity, and large biomass accumulation at 42 days after sowing (Saito et al. 2010). Dingkuhn et al. (1999) also considered relative yield (the ratio of grain yield under weedy conditions and grain yield under weed-free conditions) as an indicator of weed competitiveness and Rodenburg et al. (2009) showed that longer duration and higher yield under weed-free conditions were associated with higher grain yields under weedy conditions.

5.4. Seed traits

In annual plants, and thus in many agricultural contexts, seed size can be an important determinant of success during initial stages of competition simply because larger seeds lead to greater initial growth and therefore to greater capture of available resources at the expense of other competitors (Black, 1958; Ross and Harper, 1972). Vigorous seed, expressed as early emergence and root growth contribute to cultivar competitiveness against weeds. In sweet corn, early vigor and seedling growth rate are largely attributed to endosperm phenotype, specifically, the result of starch concentration of the endosperm (Azanza et al. 1996).

In general, seed vigour within a genotype has been attributed to seed size, protein, which is in turn related to ATP production and ultimately, mitochondrial quality and quantity. Seed vigour has been positively related to both seedling vigour and final yield and can be improved to enhance crop competitive ability (Watson et al. 2002).

6. Nutrient management

A suitable nutrient management program can be an effective tool to control weeds in cropping systems. The competitive relationship between crop and weeds is highly dependent on supply and availability of nutrients (Evans et al. 2003; Di Tomaso 1995). Manipulation of soil fertility, whether using organic or inorganic amendments should be considered as an important component of long-term weed management programs and effective fertilizer management is an important component of integrated weed management systems (Blackshaw et al. 2007; DiTomaso 1995). Unfortunately, nutrients applied to soils are also available for weeds. In most farming systems, competition for N is the most important source of nutrient interference (DiTomaso 1995). Walker and Buchanan (1982) also found that of all nutrients, plant response to nitrogen (N) fertilizer is the most widely observed and the manipulation of soil N supply offers the most promise in the short term as a means by which crop–weed competitive outcomes can be influenced.

Therefore, it is important to develop fertilization strategies for crop production that enhance the competitive ability of the crop, minimize weed competition, and reduce the risk of nonpoint source pollution from nitrogen (Cathcart and Swanton 2003; DiTomaso 1995).

6.1. Aspects of nutrient management

Different aspects of nutrient management including fertilizer rate, timing and application method can be successfully manipulated to reduce weed interference in crops (Angonin et al. 1996; Blackshaw et al. 2004; Van Delden et al. 2002).

6.1.1. Application rate and timing

Weed emergence and growth in the field can be stimulated by fertilizer application rate and timing. Generally, weed growth may increase as the nitrogen application rate increases, resulting in the need for more frequent POST herbicide applications or cultivation (Sweeney et al. 2008). In sugar beet, weed emergence from sown seed increased as the N application rate at planting increased from 56 to 224 kg N ha⁻¹ (Dotzenko et al. 1969).

The timing of fertilizer application in early planted crops, such as sugar beet and corn, may especially influence the germination, emergence, and competitiveness of weeds that might otherwise remain dormant early in the growing season. In a study, when nitrogen was broadcasted in April at the time of planting, weed germination and emergence were stimulated. In contrast, nitrogen application at the time of planting in May did not influence seed germination and weed emergence because of greater N availability because of mineralization at this time of year or because seed germination has been stimulated by other environmental cues (Sweeney et al. 2008). Results of both greenhouse (Alka"mper et al. 1979) and field experiments (Davis and Liebman 2001; Dyck et al. 1995) indicate that for certain crop-weed combinations, delaying soil N availability can shift the competitive balance to favor crop growth. For example, the competitiveness of wild mustard (*Sinapis arvensis* L.), a winter annual, in sugar beet was favored by early compared with late nitrogen fertilization (Paolini et al. 1999). Generally, delaying nitrogen applications, applying slow-release nitrogen fertilizers or placing nitrogen below the weed seed germination zone could be potential strategies for reducing early season weed establishment in cropping systems (Sweeney et al. 2008).

In the case of phosphorus, early-season application is critically important for vigorous plant growth and development (Grant et al. 2001). Thus, fertilization strategies that restrict weed access to phosphorus fertilizer early in the growing season would appear to have merit (Blackshaw and Molnar 2009).

6.1.2. Application method

Crop-weed interactions can also influence by fertilizer application method. Since, weeds often germinate at or near the soil surface, especially in zero-tillage systems (Yenish et al., 1992; Hoffman et al., 1998), therefore, in this situation the greatest benefits may be realized by physically placing nitrogen (N) in an area of the soil profile where crop seeds, but not weed seeds, are germinating (Blackshaw 2005). Subsurface-banded N was often better than surface-broadcast N fertilizer in terms of N uptake by wheat vs. weeds, weed biomass production and wheat yield (Blackshaw et al. 2005).

Petersen (2003) reported that weed N uptake and weed biomass were 50% lower with subsurface-banded compared with surface-broadcast liquid swine manure. Rasmussen (2002) similarly documented lower weed biomass and higher crop yield with injected than with surface-broadcast liquid swine manure. In another study, subsurface-banded N compared with broadcast N fertilizer reduces N uptake by weeds and decreases weed growth and biomass. Moreover, banded N fertilizer resulted in the lowest seed bank numbers of both grass and broadleaf weeds (Blackshaw 2005).

Phosphorus (P) fertilization practices could also have an impact on the extent of interference by weeds. Researchers indicated that banding P near lettuce rooting system as opposed to broadcast P could potentially reduce the damage of spiny amaranth (*Amaranthus spinosus*), smooth pigweed, and common purslane by enhancing the competitive ability of the crop (Santos et al. 1997; Shrefler et al. 1994). In wheat, seed-placed or mid row-banded P compared with surface-broadcast P fertilizer often resulted in higher yields when wheat was in the presence of competitive weeds (Blackshaw and Molnar 2009).

In general, weed P concentration and biomass production were often greatest with surface-broadcast P fertilizer, indicating that this common application method of P fertilizer should be discouraged. Alternative practices such as seed-placed or subsurface-banded P fertilizer were less advantageous to weeds. Weed seed bank was also affected by P application method. So that, seedbank evaluation at the end of the experiment indicated that the seed density of five of six weed species under study was reduced with seed-placed or subsurface-banded P compared with surface-broadcast P (Blackshaw and Molnar 2009).

However, the benefit of seed-placed or subsurface-banded P fertilizer will likely be greatest in soils with low background P levels and within zero-tillage production systems, where weed seeds are not distributed throughout the soil profile but rather concentrated near the soil surface (Blackshaw and Molnar 2009).

6.2. Organic and biofertilizer

Organic manure may affect crop–weed competitive interactions differently than chemical nitrogen fertilizer (Davis and Liebman, 2001), probably due to speed of N release or form of N. In a study, the gradual N release from manure and compost over years appeared to benefit weeds more than spring wheat. Moreover, fresh and composted manure had the greatest seed bank of both grass and broadleaf species as compared with chemical fertilizer treatment (Blackshaw 2005). In another study, Mohammadi et al. (2012) found that phosphate biofertilizer had no significant effect on corn yield, whereas, weed biomass was notably increased when phosphate biofertilizer was applied.

It seems that, in most cases, weed infestation level and duration may enhance by the use of organic fertilizers. However, the other beneficial aspects of these fertilizers should not be ignored.

6.3. Critical period of weed control in response to nutrient management

The critical period of weed control (CPWC) is an important principal of an integrated weed management program. It is a period in the crop growth cycle during which weeds must be controlled to prevent yield losses (Knezevic et al. 2002). Weeds that are present before or emerge after this period do not cause significant yield loss. Studies on the critical period of weed control are important in making weed control recommendations because they indicate the optimum time for implementing and maintaining weed control and reduce cost of weed control practices (Hall et al. 1992; Van Acker et al. 1993).

According to Weaver et al. (1992) the manipulation of edaphic factors including the alteration of soil nutrient supply can influence the crop-weed interference relationships, especially in determining the critical time of weed removal (the start of the critical period). Evans et al. (2003) reported that the addition of nitrogen fertilizer delayed the beginning and hastened the end of the critical period of weed control in corn. Their study showed that the effect of nitrogen fertilization on early season crop growth provided a competitive advantage for corn relative to weeds. In another study, Mohammadi and Amiri (2011) found that the use of mono ammonium phosphate as a starter fertilizer slightly delayed the end of the CPWC in soybean (by 5 days), but this condition shortened the CPWC by 12 days because of the later beginning of the CPWC (by 17 days).

6.4. Weed response to fertilization

Some weed species are considered to be luxury consumers of nutrients (Qasem 1992; Teyker et al. 1991) and this might contribute to their ability to take up higher amounts of N at higher N fertilizer rates. Weeds not only reduce the amount of N available to crops but the growth of many weed species is enhanced by higher soil N levels (Blackshaw et al. 2003; Henson and Jordan 1982; Supasilapa et al. 1992). Thus, adding N fertilizer in cropping systems can potentially have the unintended consequence of increasing the growth and competitive ability of weeds more than that of the crop.

In a greenhouse study, Teyker et al. (1991) reported greater N uptake for redroot pigweed (*Amaranthus retroflexus* L.) than corn when the addition of N was elevated, suggesting that redroot pigweed interference in corn may be greater at higher levels of N. Others also have postulated that weeds may be more competitive when fertility is enhanced with N addition because of the superior uptake efficiency of many weed species (Di Tomaso 1995; Sibuga and Bandeen 1980).

At the conclusion of a 47-yr soil fertility study, carpetweed (*Mollugo verticillata* L.) and henbit (*Lamium amplexicaule* L.) densities were greatest on plots that had received annual applications of P fertilizer (Banks et al. 1976). In another study, downy brome (*Bromus tectorum* L.) densities were reported to be higher on soils with higher P levels (Belnap et al. 2003). Verma et al. (1999) similarly reported that weed growth and competitiveness with fenugreek (*Trigonella foenum-graecum* L.) increased at the higher soil P levels.

Weed germination and dormancy are also influenced by fertilizer application. For example, germination of common lambsquarters seed from mother plants that received 280 kg ha^{-1} of ammonium nitrate was greater than germination of seed from a mother plant where no N was applied, suggesting that N deficiency increased dormancy in seeds (Baskin and Baskin 1998; Fawcett and Slife 1978). According to Cairns and de Villiers (1986) the dormancy of several grass weed species was broken by ammonia, but the gas had no effect on the dormancy of dicotyledonous weed seed. Redroot pigweed (*Amaranthus retroflexus* L.) seed germination was also stimulated by 10 to 100 ppmv of ammonium nitrate or urea (Sardi and

Beres 1996). Other researchers found that the chilling or light requirement for seed germination in some species can be replaced with N, particularly nitrate (Cohn et al. 1983; Egley and Duke 1985; Sexsmith and Pittman 1963; Steinbauer and Grigsby 1957).

Generally, fertilizer management strategies that favour crops over weeds deserve greater attention when weed infestations consist of species known to be highly responsive to higher soil nutrient (e.g. N) levels. In these situations farmers should consider the benefits of specific fertilizer timing and/or placement methods that would minimize weed interference (Blackshaw et al. 2004; Kirkland and Beckie 1998; Mesbah and Miller 1999).

7. Biological control

Biological control of weeds refers to the use of any kind of organism (micro or macro) to suppress weeds and reduce their harmful effects in agroecosystems. Plant pathogens are potentially valuable additions to the arsenal of weapons for use against weeds.

7.1. Weed biological control approaches using pathogens

Biological control of weeds using pathogens can be considered from two broad approaches including classical biological control (CBC) and inundative biological control (IBC) (Mohan babu et al. 2003).

7.1.1. Classical biological control

This approach is fairly simple in its concept: discover effective and highly host-specific agents from the weed's native geographic range, confirm their safety and effectiveness by rigorous experimental evaluation, and introduce them into regions where the weed has been newly introduced and requires control (Charudattan and Dinoor 2000). Host specificity tests provide the information on which to base the risk assessment and, thereby play the central role in any CBC project (Mohan babu et al. 2003).

Classical biological control by means of pathogens has been used in several parts of the world to control exotic weeds (Bruckart and Hasan 1991; Watson 1991). One of the most successful examples of classical biological control of weeds is the introduction of a rust fungus, *Puccinia chondrillina*, into Australia to control rush skeleton weed (*Chondrilla juncea*). A plant of Mediterranean origin, it became a serious weed in Australian cereal crops. The fungus, also from the Mediterranean was introduced along with three insects, as a classical biocontrol agent. Following the introduction and establishment, the fungus disseminated rapidly and widely and controlled the most common biotype of the weed (Cullen, 1985).

Other successful examples of classical biocontrol programs include the use of a smut fungus, *Entyloma ageratinae*, imported from Jamaica to control Hamakua pamakani (*Ageratina riparia*, Asteraceae) in Hawaiian forests and rangelands (Trujillo, 1985) and three other rust fungi, *Puccinia carduorum*, imported from Turkey and released into northeastern United States to control musk thistle (*Carduus thoermeri*) (Baudoin et al., 1993), *Phragmidium violaceum* to control weedy species of *Rubus* in Chile (Oehrens, 1977) and Australia (Bruzzese, 1995) and *Uromycladium tepperianum*, to control an introduced invasive tree species, *Acacia saligna* in South Africa (Morris, 1997). The last fungus causes extensive gall formation on branches and twigs accompanied by a significant energy loss. Heavily infected trees are eventually killed (Charudattan and Dinoor 2000).

However, there is a potential problem in using biological control, namely, a shift in the weed population toward more resistant weed biotypes. Although, it also illustrates the possibility to counter the presence of natural resistance in weed populations by the introduction of new pathogen strains (Charudattan and Dinoor 2000).

7.1.2. Inundative biological control

The strategy of inundative biological control is to simulate natural epiphytotics of a selected pathogen within the population of the target weed species, early in the season and thus kill or at least significantly reduce the competitive ability of the weed and so prevent crop losses. This approach is typically used against endemic weeds, in which indigenous pathogens are mass-produced and applied as formulated products (bioherbicides) (Mohan babu et al. 2003). A bioherbicide is defined as a plant pathogen used as a weed-control agent through inundative and repeated applications of its inoculum (Charudattan and Dinoor 2000). The specificity of bioherbicides is considered as a positive attribute (Mohan babu et al. 2003).

Some examples of registered bioherbicides consisted of DeVine®, composed of a Florida isolate of *Phytophthora palmivora*, is used for the control of *Morrenia odorata* (stranglervine or milkweed vine) in citrus in Florida. Collego®, based on *Colletotrichum gloeosporioides* f.sp. *aeschynomene*, is used to control *Aeschynomene virginica* (northern jointvetch), a leguminous weed in rice and soybean crops in Arkansas, Mississippi and Louisiana. BioMal®, registered in Canada for the control of *Malva pusilla* (round-leaved mallow), containing *Colletotrichum gloeosporioides* f.sp. *malvae*, is presently unavailable for commercial use. The fourth bioherbicide, Dr. BioSedge®, based on the rust fungus *Puccinia canaliculata* and registered for the control of *Cyperus esculentus* (yellow nutsedge) in the United States, is also unavailable for commercial use. An isolate of *Xanthomonas campestris* pv. *poae*, a wilt-inducing bacterium, isolated in Japan from *Poa annua* (annual bluegrass or wintergrass), is registered in Japan as the bioherbicide CAMPERICO® to control annual bluegrass in golf courses (Charudattan and Dinoor 2000).

Bioherbicides can make a significant contribution to weed control in the future, once the well-documented constraints have been overcome, particularly through improved target selection, formulation and marketing. The over-riding concern in using plant pathogens for weed control is their potential threat to non-targets (Mohan babu et al. 2003) which needs a serious attention.

7.2. Microbial-derived herbicides

Microbial-derived herbicides, especially microorganism secondary metabolites are a new kind of microbial herbicide to control weeds which are always phytotoxins. They are very different in chemical structure and size. These bioactive components invade into the host plant, cause pathogenicity, destroy their structure and lead them to produce necrotic lesions or chlorotic halo (Li et al. 2003). They are less poisonous to most of mammalian systems, easily degraded and so far result in no biological disaster compared to chemical herbicides (Charudattan, 1991).

Phytotoxins used for microbial herbicides can be divided into three types: bacterial, fungal and actinomycete derived product. The pathogens which produce phytotoxins as a microbial herbicide must fit certain requirements: (1) be reproduced by biological technique, (2) grow fast after spraying or be capable of killing weeds within definite time, (3) suit industrial production and (4) be suitable for packaging, transport and use (Li et al. 2003).

In comparison with fungi, bacteria have some advantageous characteristics such as short-growth period, simple fermentation technique and easily controlled production process. In addition, bacteria can produce secondary metabolites unlike fungal spores, which need strict conditions for action as herbicides and their residues are easily degraded. Bacterial herbicides have a good prospect in application and exploitation (Li et al. 2003). Most of the bacteria with an ability to produce toxins are Gram-positive bacteria such as *Pseudomonas, Erwinia, Xanthomonas* but there are a few Gram-negative bacteria such as *Streptomyces, Corynebacterium fasciomonads* and some are non-fluorescent *pseudomonads* (Kremer et al., 1990).

Two phytotoxins from actinomycetes including herbicidines and herbimycins are higher-plant toxins and produced by *Streptomyces saganonensis*. The former is used to control grassy weeds in paddy field as a selective herbicide, the latter controls monocotyledonous and dicotyledonous weeds (Stephen and Lydon, 1987).

However, the role of biomicrobial herbicides in agriculture is still problematic and insignificant (Mohan babu et al. 2003). Although, the current emphasis on lowering use of chemical herbicides may increase the production and use of biological-based herbicides in the future.

In general, the use of alternative weed control strategies can prevent or reduce the chemical herbicide application. This can lead to less reliance on fossil fuels (the non-renewable energy resources), lower environmental degradations and consequently a higher degree of sustainability for agroecosystems.

Author details

G.R. Mohammadi

Department of Crop Production and Breeding, Faculty of Agriculture and Natural Resources, Razi University, Kermanshah, Iran

References

[1] Aarssen, L.W., 1992. Causes and consequences of variation in competitive ability in plant communities. J. Veg. Sci. 3, 165–174.

[2] Akanvou, R., L. Bastiaans, M. J. Kropff, J. Goudriaan and M. Becker. 2001. Characterization of Growth, Nitrogen Accumulation and Competitive Ability of Six Tropical Legumes for Potential Use in Intercropping Systems. Journal of Agronomy and Crop Science 187, 111-120.

[3] Akemo, M.C., Regnier, E.E. and Bennett, M.A. 2000. Weed suppression in spring-sown rye-pea cover crop mixes. Weed Technology 14, 545-549.

[4] Alcorta M., Fidelibus M. W., Steenwerth K. L. and Shrestha A. 2011. Effect of Vineyard Row Orientation on Growth and Phenology of Glyphosate-Resistant and Glyphosate-Susceptible Horseweed (*Conyza canadensis*). Weed Sc. 59(1):55-60.

[5] Alka¨mper, J., E. Pessips, and D. V. Long. 1979. Einfluss der Du°ngung auf die Entwickelung und Nahrstoffnahme verschiedener Unkra°uter in Mais. European Weed Research Society Symposium, Mainz, Germany. Paris: European Weed Research Society.

[6] Angiras, N. and V. Sharma. 1996. Influence of row orientation, row spacing and weed-control methods on physiological performance of irrigated wheat (*Triticum aestivum*). Indian J. Agron. 41:41–47.

[7] Angonin, C., J. P. Caussanel, and J. M. Meynard. 1996. Competition between winter wheat and *Veronica hederifolia*: influence of weed density and the amount and timing of nitrogen application. Weed Res. 36:175–187.

[8] Azanza, F., A. B. Zur, and J. Juvik. 1996. Variation in sweet corn characteristics associated with stand establishment and eating quality. Euphytica 87:7–18.

[9] Ball, D. A., Ogg, Jr., A. G. and Chevalier, P. M. 1997. The influence of seeding rate on weed control in small-red lentil (*Lens culinaris*). Weed Sci. 45: 296-300.

[10] Banks, P. A., P. W. Santelmann, and B. B. Tucker. 1976. Influence of long-term soil fertility treatments on weed species in winter wheat. Agron. J. 68:825-827.

[11] Barberi, P., 2002. Weed management in organic agriculture: are we addressing the right issues? Weed Res. 42, 177–193.

[12] Barnes, J. P., Putnam, A. R. and Burke, B. A. 1985. Allelopathic activity of rye (*Secale cereale* L.). In The Science of allelopathy (ed. A. R. Putnam and C. S. Tang), pp. 271-286. Wiley, New York.

[13] Barnes, J.P., Putnam, A.R., 1983. Rye residues contribute to weed suppression in no–tillage cropping systems. J. Chem. Ecol. 9, 1045–1057.

[14] Barton, D. L., Thill, D. C. and Shafii, B. 1992. Integrated wild oat (*Avena fatua*) management affects spring barley (*Hordeum vulgare*) yield and economics. Weed Technol. 6: 129–135.

[15] Baskin, C. C. and J. M. Baskin. 1998. Seeds: Ecology, Biogeography, and Evolution of Dormancy and Germination. New York: Academic.

[16] Baudoin, A.B.A.M., Abad, R.G., Kok, L.T., Bruckart, W.L., 1993. Field evaluation of *Puccinia carduorum* for biological control of musk thistle. Biol. Control 3, 53–60.

[17] Begna, S. H., R. I. Hamilton, L. M. Dwyer, D. W. Stewart, D. Cloutier, L. Assemat, K. Foroutan-Pour, and D. L. Smith. 2001a. Weed biomass production response to plant spacing and corn (*Zea mays*) hybrids differing in canopy architecture. Weed Technol. 15:647–653.

[18] Begna, S. H., D. L. Smith, R. I. Hamilton, L. M. Dwyer, and D. W. Stewart. 2001b. Corn genotypic variation effects on seedling emergence and leaf appearance of short-season areas. J. Agro. Crop Sci. 186:267–271.

[19] Belnap, J., S. K. Sherrod, and M. E. Miller. 2003. Effects of soil amendments on germination and emergence of downy brome (*Bromus tectorum*) and *Hilaria jamesii*. Weed Sci. 51:371–378.

[20] Bennett, A. C. and D. R. Shaw. 2000. Effect of Glycine max cultivar and weed control on weed seed characteristics. Weed Sci. 48:431–435.

[21] Berkowitz, A. R. 1988. Competition for resources in weed-crop mixtures. Pages 89–119 in M. A. Altieri and M. Liebman, eds. Weed management in agroecosystems: Ecological approaches. CRC Press, Inc., Boca Raton, FL.

[22] Bhowmik P. C. and Inderjit. 2003. Challenges and opportunities in implementing allelopathy for natural weed management. Crop Prot 22: 661-671.

[23] Bhowmik, P.C., 1988. Cinmethylin for weed control in soybeans, *Glycine max*. Weed Sci. 36, 678–682.

[24] Black, J.N., 1958. Competition between plants of different initial seed sizes in swards of subterranean clover (*Trifolium subterraneum* L.) with particular reference to leaf area and light microclimate. Aust J Agric Res 9: 299–318.

[25] Blackshaw R. E. 2005. Nitrogen fertilizer, manure, and compost effects on weed growth and competition with spring wheat. Agron. J. 97:1612–1621.

[26] Blackshaw R. E. and Molnar L. J. 2009. Phosphorus fertilizer application method affects weed growth and competition with wheat. Weed Sci. 57:311-318.

[27] Blackshaw R. E., Molnar L. J. and Larney F. J. 2005. Fertilizer, manure and compost effects on weed growth and competition with winter wheat in western Canada. Crop Protection 24: 971–980.

[28] Blackshaw, R. E., L. J. Molnar, and H. H. Janzen. 2004. Nitrogen fertilizer timing and application method affect weed growth and competition with spring wheat. Weed Sci. 52: 614–622.

[29] Blackshaw, R. E., R. L. Anderson, and D. Lemerle. 2007. Cultural weed management. Pages 35–47 in M. K. Upadhyaya and R. E. Blackshaw, eds. Non-Chemical Weed Management: Principles, Concepts and Technology. Oxfordfordshire, UK: CABI.

[30] Blackshaw, R.E., 1994. Differential competitive ability of winter wheat cultivars against Downy brome. Agron. J. 86, 649–654.

[31] Blackshaw, R.E., Brandt, R.N., Janzen, H.H., Entz, T., Grant, C.A., Derksen, D.A., 2003. Differential response of weed species to added nitrogen. Weed Sci. 51: 532–539.

[32] Bond W. and Grundy A. C. 2001. Non-chemical weed management in organic farming systems. Weed Res., 41, 383-405.

[33] Borger C. P. D., Hashem A. and Pathan S. 2010. Manipulating crop row orientation to suppress weeds and increase crop yield. Weed Sci. 58:174-178.

[34] Börner. 1960. Liberation of organic substances from higher plants and their role in the soil sickness problem. The Botanical Review 26, 396-424.

[35] Boydston, R.A., Hang, A., 1995. Rapeseed (*Brassica napus*) green manure crop suppresses weeds in potato (*Solanum tuberosum*). Weed Technol. 9, 669–675.

[36] Brainard, D.C. and Bellinder, R.R. 2004. Weed suppression in a broccoli-winter rye intercropping system. Weed Science 52, 281-290.

[37] Brainard, D.C., Bellinder, R.R., Miller, A.J., 2004. Cultivation and interseeding for weed control in transplanted cabbage. Weed Technol. 18, 704–710.

[38] Brandsaeter, L.O., Netland, J., 1999. Winter annual legumes for use as cover crops in row crops in northern regions: I. Field experiments. Crop Sci. 39, 1369–1379.

[39] Bruckart, W.L., Hasan, S., 1991. Options with plant pathogens intended for classical control of range and pasture weeds. In: TeBeest, D.O. (Ed.), Microbial Control of Weeds. Chapman and Hall, New York, pp. 69-79.

[40] Bruzzese, E., 1995. Present status of biological control of European blackberry (*Rubus fruticosus* aggregate in Australia. In: Delfosse, E.S., Scott, R.R. (Eds.), Proceedings of the Eighth International Symposium on Biological Control of Weeds. DSIR/CSIRO. Melbourne, Australia, pp. 297-299.

[41] Bussan, A.J., O.C. Burnside, J.H. Orf, E.A. Ristau, and K.J. Puettmann. 1997. Field evaluation of soybean (*Glycine max*) genotypes for weed competitiveness. Weed Sci. 45:31-37.

[42] Cairns, A.L.P. and O. T. de Villiers. 1986. Breaking dormancy of *Avena fatua* L. seed by treatment with ammonia. Weed Res. 26:191–198.

[43] Callaway, M.B. 1992. A compendium of crop varietal tolerance to weeds. Am. J. Alternative Agric. 7:169–180.

[44] Callaway, M.B., and F. Forcella. 1993. Crop tolerance to weeds. p. 100–131. *In* C.A. Francis and M.B. Callaway (ed.) Crop improvement for sustainable agriculture. Univ. Nebraska Press, Lincoln.

[45] Campiglia E., Paolini R., Colla G. and Mancinelli R. 2009. The effects of cover cropping on yield and weed control of potato in a transitional system. Field Crops Research 112, 16–23.

[46] Carmona, D. M. and D. A. Landis. 1999. Influence of refuge habitats and cover crops on seasonal-density of ground beetles (Coleoptera: Carabidae) in field crops. Environ. Entomol. 28:1145–1153.

[47] Cathcart, R. J. and C. J. Swanton. 2003. Nitrogen management will influence threshold values of green foxtail (*Setaria viridis*) in corn. Weed Sci. 51:975–986.

[48] Charudattan, 1991. The mycoherbicide approach with plant pathogens. In: TeBeest, D.O. (Ed.), Microbial Control of Weeds. Chapman and Hall, New York, pp. 24–57.

[49] Charudattan, R., Dinoor, A., 2000. Biological control of weeds using plant pathogens: accomplishments and limitations. Crop Prot. 19, 691–695.

[50] Chauhan B. S. and Johnson D. E. 2011. Row spacing and weed control timing affect yield of aerobic rice. Field Crops Res. 121:226-231.

[51] Chou, C. H. 1999. Roles of allelopathy in plant biodiversity and sustainable agriculture. Critical Rev. Plant Sci. 18:609–636.

[52] Cohn, M. A., D. L. Butera, and J. A. Hughs. 1983. Seed dormancy in red rice, III: response to nitrite, nitrate, and ammonium ions. Plant Physiol 73:381–384.

[53] Collins A. S., Chase C. A., Stall W. M. and Hutchinson C. M. 2008. Optimum Densities of Three Leguminous Cover Crops for Suppression of Smooth Pigweed (*Amaranthus hybridus*). Weed Science 56, 753–761.

[54] Conklin, A. E., M. S. Erich, M. Liebman, D. Lambert, E. R. Gallandt, and W. A. Halteman. 2002. Effects of red clover (*Trifolium pratense*) green manure and compost soil amendments on wild mustard (*Brassica kaber*) growth and incidence of disease. Plant Soil 238:245-256.

[55] Copaja, S. V., Nicol, D. and Wratten, S. D. 1999. Accumulation of hydroxamic acids during wheat germination. Phytochemistry 50, 17-24.

[56] Cousens, R., and Mortimer, M. 1995. Dynamics of Weed Populations. Cambridge University Press, Cambridge.

[57] Creamer, N. G., Bennett, M. A., Stinner, B. R., Cardina, J. and Regnier, E. E. 1996. Mechanisms of weed suppression in cover crop-based production systems. Hort science 31, 410-413.

[58] Cromar, H. E., Murphy, S. D., and Swanton, C. J. 1999. Influence of tillage and crop residue on post dispersal predation of weed seeds. Weed Science, 47, 184-194.

[59] Crotser, M. P. and W. W. Witt. 2000. Effect of *Glycine max* canopy characteristics, *G. max* interference, and weed free period on *Solanum ptycanthum* growth. Weed Sci. 48:20–26.

[60] Cullen, J.M., 1985. Bringing the cost benefit analysis of biological control of *Chondrilla juncea* up to date. In: Delfosse, E.S. (Ed.), Proceedings of the VI International Symposium on Biological Control of Weeds. Agriculture Canada, Ottawa, pp. 145-152.

[61] Czarnota, M.A., A.M. Rimando, and L.A. Weston. 2003. Evaluation of root exudates of seven sorghum accessions. J. Chem. Ecol. 29: 2073–2083.

[62] Daferera, D. J., B. N. Ziogas, and M. G. Polissiou. 2003. The effectiveness of plant essential oils on the growth of *Botrytis cinerea, Fusarium sp.* and *Clevibacter michiganensis* subsp. michiganensis. Crop Prot. 22:39–44.

[63] Daughtry, C.S.T., Gallo, K.P., Bauer, M.E., 1983. Spectral estimates of solar radiation intercepted by corn canopies. Agron. J. 72, 527–531.

[64] Davis, A. S. and M. Liebman. 2001. Nitrogen source influences wild mustard growth and competitive effect on sweet corn. Weed Sci. 49:558–566.

[65] Davis, A. S. and M. Liebman. 2003. Cropping system effects on *Setaria faberi* seed bank dynamics. Aspects Appl. Biol. 69:83–91.

[66] Dayan, F.E. 2002. Natural pesticides. p. 521–525. In D. Pimentel (ed.) Encyclopedia of pest management. Marcel Dekker, Inc., New York.

[67] Dayan, F.E., J.G. Romagni, M. Tellez, A. Rimando, and S. Duke. 1999. Managing weeds with bio-synthesized products. Pestic. Outl. 10: 185–188.

[68] De Haan R. L., D. L. Wyse, N. J. Ehlke, B. D. Maxwell, and D. H. Putnam. 1994. Simulation of spring-seeded smoother plants for weed control in corn (*Zea mays*). Weed Sci. 42,35–43.

[69] den Hollander N. G., Bastiaans L. and Kropff M. J. 2007. Clover as a cover crop for weed suppression in an intercropping design I. Characteristics of several clover species. Europ. J. Agronomy 26: 92–103.

[70] Dhima, K. V., I. B. Vasilakoglou, Th. D. Gatsis, E. Panou-Philotheou, and I. G. Eleftherohorinos. 2009. Effects of aromatic plants incorporated as green manure on weed and maize development. Field Crops Res. 110:235–241.

[71] Di Tomaso JM 1995 Approaches for improving crop competitiveness through the manipulation of fertilization strategies. Weed Sci 43: 491-497.

[72] Dilday, R. H., Lin, J. and Yan, W. 1994. Identification of allelopathy in the Usda-Ars rice germplasm collection. Australian Journal of Experimental Agriculture 34, 907-910.

[73] Dilday, R.H., Mattice, J.D., Moldenhauer, K., Yan, AW., 2001. Allelopathic potential in rice germplasm against ducksalad, redstem and barnyard grass. J. Crop Prod. 4, 287–301.

[74] Dingkuhn M., Johnson D.E., Sow A. and Audebert A.Y. 1999. Relationships between upland rice canopy characteristics and weed competitiveness. Field Crops Res. 61, 79–95.

[75] Dotzenko, A. D., M. Ozkan, and K. R. Storer. 1969. Influence of crop sequence, nitrogen fertilizer and herbicides on weed seed populations in sugar beet fields. Agron. J. 61:34–37.

[76] Drews S., Juroszek P., Neuhoff D., Kopke U. 2004. Optimizing shading ability of winter wheat stands as a method of weed control. J. Plant Dis. Prot.:545–552.

[77] Dudai, N., A. Poljakoff-Mayber, A. M. Mayer, E. Putievsky, and H. R. Lerner. 1999. Essential oils as allelochemicals and their potential use as bioherbicides. J. Chem. Ecol. 25:1079–1089.

[78] Duke, S.O., Dayan, F.E., Ramagni, J.G., Rimando, A.M., 2000. Natural products as sources of herbicides: current status and future trends. Weed Res. 40: 99–111.

[79] Duke, S.O., Dayan, F.E., Rimando, R.M., Schrader, K.K., Aliotta, G., Oliva, A., Romagni, J.G., 2002. Chemicals from nature for weed management. Weed Sci. 50, 138–151.

[80] Duke, S.O., Vaughn, K.C., Croom, E.M., Elsohly, H.N., 1987. Artemisinin, a constituent of annual wormwood (*Artemisia annua*) is a selective phytotoxin. Weed Sci. 35, 499–505.

[81] Dyck, E. and M. Liebman. 1994. Soil fertility management as a factor in weed control: the effect of crimson clover residue, synthetic nitrogen fertilizer, and their interaction on emergence and early growth of lambsquarters and sweet corn. Plant Soil 167:227–237.

[82] Dyck, E., M. Liebman, and M. S. Erich. 1995. Crop-weed interference as influenced by a leguminous or synthetic fertilizer nitrogen source: I. Double cropping experiments with crimson clover, sweet corn, and lambsquarters. Agric. Ecosyst. Environ. 56:93–108.

[83] Egley, G. H. and S. O. Duke. 1985. Physiology of weed seed dormancy and germination. Pages 27–64 in Weed Physiology: Reproduction and Ecophysiology, Volume 1. Boca Raton, FL: CRC.

[84] Einhellig, F.A., and J.A. Rasmussen. 1989. Prior cropping with grain sorghum inhibits weeds. J. Chem. Ecol. 15:951–960.

[85] Entz, M.H., W.J. Bullied, and F. Katepa-Mupondwa. 1995. Rotational benefits of forage crops in Canadian prairie cropping systems. J. Prod. Agric. 8:521-529.

[86] Evans S.P., Knezevic S.Z., Shapiro C., Lindquist J.L. 2003. Nitrogen level affects critical period for weed control in corn. Weed Sci 51: 408-417.

[87] Facelli, J. M., and S. T. A. Pickett. 1991. Plant litter, It's dynamics and effects on plant community structure. Bot. Rev. 57,1–32.

[88] Fawcett, R. S. and F. W. Slife. 1978. Effects of field applications of nitrate on weed seed germination and dormancy. Weed Sci. 26:594–596.

[89] Fay, P. K. and Duke, W. B. 1977. An assessment of allelopathic potential in *Avena* germplasm. Weed Science 25, 224-228.

[90] Fujii, Y. 1999. Allelopathy of hairy vetch and *Macuna*; their application for sustainable agriculture. pp.289-300. *In* C.H. Chou *et al. Biodiversity and Allelopathy from Organisms to Ecosystems in the Pacific.* Academia Sinica, Taipei.

[91] Gallandt, E. R., T. Molloy, R. P. Lynch, and F. A. Drummond. 2005. Effect of cover-cropping systems on invertebrate seed predation. Weed Sci. 53:69–76.

[92] Garrity, D.P., M. Movillon, and K. Moody. 1992. Differential weed suppression ability in upland rice cultivars. Agron. J. 84:586–591.

[93] Gaudet, C.L., Keddy, P.A., 1988. A comparative approach to predicting competitive ability from plant traits. Nature 334, 242–243.

[94] Gianoli, E. and Niemayer, H. M. 1998. DIBOA in wild Poaceae: Source of resistance to the Russian wheat aphid (*Diuraphis noxia*) and the greenbug (*Schizaphis graminum*). Euphytica 102, 317-321.

[95] Girardin, Ph. and Tollenaar, M. 1994. Effects of intraspecific interference on maize leaf azimuth. Crop Sci. 34: 151–155.

[96] Goldberg, D.E., 1990. Components of resource competition in plant communities. In: Grace, J.B., Tilman, D. (Eds.), Perspectives on Plant Competition. Academic Press, San Diego, pp. 27–49.

[97] Gonzalez, V.M., Kazimir, J., Nimbal, C., Weston, L.A., Cheniae, G.M., 1997. Inhibition of a photosystem II electron transfer reaction by the natural product sorgoleone. J. Agric. Food Chem. 45, 1415–1421.

[98] Gooding, M.J., A.J. Thompson, and W.P. Davies. 1993. Interception of photosynthetically active radiation, competitive ability, and yield of organically grown wheat varieties. Aspects Appl. Biol. 34:355–362.

[99] Grace J.B. 1990. On the relationship between plant traits and competitive ability. In: Perspective on Plant Competition (ed. by Grace J.B. and Tilman D.). Academic Press, San Diego, CA, 51–65.

[100] Grant, C. A., D. N. Flaten, D. J. Tomasiewicz, and S. C. Sheppard. 2001. The importance of early season phosphorus nutrition. Can. J. Plant Sci. 81:211–224.

[101] Grime J.P. 1977. Evidence for the existence of three primary strategies in plants and its relevance to ecological and evolutionary theory. *Am. Nat.* 111, 1169–1194.

[102] Hall M. R., Swanton C. J. and Anderson G. W. 1992. The critical period of weed control in grain corn (*Zea mays*). Weed Sci 40: 441-447.

[103] Hallberg G. R. 1989. Pesticide pollution of ground water in the humid United States. Agric. Ecosyst. Environ. 26, 299–367.

[104] Harker, K.N., J.T. O'Donovan, R.B. Irvine, T.K. Turkington, and G.W. Clayton. 2009. Integrating cropping systems with cultural techniques augments wild oat (*Avena fatua*) management in barley. Weed Sci. 57:326–337.

[105] He, H.Q., L.H. Shen, J. Xiong, X.L. Jia, W.X. Lin, and H. Wu. 2004. Conditional genetic effect of allelopathy in rice (*Oryza sativa* L.) under different environmental conditions. Plant Grow. Reg. 44: 211–218.

[106] Henson, J. F. and Jordan, L. S. 1982. Wild oat (*Avena fatua*) competition with wheat (*Triticum aestivum* and *T. turgidum* Durum) for nitrate. Weed Sci., 30: 297-300.

[107] Hinson,K., andW.D.Hanson. 1962. Competition studies on soybeans. Crop Sci. 2:117-123.

[108] Hoffman, M.L., M.D.K. Owen, and D.D. Buhler. 1998. Effects of crop and weed management on density and vertical distribution of weed seeds in soil. Agron. J. 90:793-799.

[109] Holt J. S. and LeBaron H. N. 1990. Significance and distribution of herbicide resistance. Weed Technol. 4, 141-149.

[110] Holt J.S. and Ocrutt D.R. 1991. Functional relationships of growth and competitiveness in perennial weeds and cotton (*Gossypium hirsutum*). Weed Sci. 39, 575–584.

[111] Holt, J. S. 1995. Plant responses to light: a potential tool for weed management. Weed Sci. 43:474–482.

[112] Hucl, P. 1998. Response to weed control by four spring wheat genotypes differing in competitive ability. Can. J. Plant Sci. 78:171–173.

[113] Huel, D.G. and P. Hucl, 1996. Genotypic variation for competitive ability in spring wheat. Plant Breed 115: 325–329.

[114] Hunt, R., and J.H.C. Cornelissen. 1997. Components of relative growth rate and their interrelations in 59 plant species. New Phytol. 135:395–417.

[115] Inderjit and Olofsdotter M. 1998. Using and improving laboratory bioassays in rice allelopathy research. In Allelopathy in Rice. Proceedings of the workshop on Allelopathy in Rice, 25-27 Nov. 1996 (ed. M. Olofsdotter), pp. 45-55. IRRI, Manilla.

[116] Isman, B. M. 2000. Plant essential oils for pest and disease management. Crop Prot. 19:603–608.

[117] Jannink, J. L., N. R. Jordan, and J. H. Orf. 2001. Feasibility of section for high weed suppressive ability in soybean: absence of tradeoffs between rapid initial growth and sustained later growth. Euphytica 120:291–300.

[118] Jannink, J.L., Orf, J.H., Jordan, N.R., Shaw, R.G., 2000. Index selection for weed suppressive ability in soybean. Crop Sci. 40, 1087–1094.

[119] Jensen, L.B., Courtois, B., Shen, L., Li, Z., Olofsdotter, M., Mouleon, R.D. 2001. Locating genes controlling allelopathic effect against barnyardgrass in upland rice. Agron. J. 93: 21–26.

[120] Johnson D.E., Ding Kuhn M., Jones M.P. and Mahamane M.C. 1998. The influence of rice plant type on the effect of weed competition on *Oryza sativa* and *Oryza glaberrima*.Weed Res. 38: 207–216.

[121] Jordan N. 1993. Prospects for weed control through interference. Ecol. Appl. 3, 84–91.

[122] Keddy, P.A., Shipley, B., 1989. Competitive hierarchies in herbaceous plant communities. Oikos 54: 234–241.

[123] Khan, M., Donald, W. W. and Prato, T. 1996. Spring wheat (*Triticum aestivum*) management can substitute for diclofop for foxtail (*Setaria* spp.) control. Weed Sci. 44: 362–372.

[124] Kirkland, K. J. 1993. Weed management in spring barley (*Hordeum vulgare*) in the absence of herbicides. J. Sustain. Agric. 3: 95–104.

[125] Kirkland, K. J. and H. J. Beckie. 1998. Contribution of nitrogen fertilizer placement to weed management in spring wheat (*Triticum aestivum*). Weed Technol. 12:507–514.

[126] Knezevic S. Z., Evans S. P., Blankenship E. E., Van Acker R. C. and Lindquist J. L. 2002. Critical period for weed control: the concept and data analysis. Weed Sci 50: 773–786.

[127] Kohli, R.K., Batish, D., Singh, H.P., 1998. Allelopathy and its implications in agroecosystems. J. Crop Prod. 1, 169–202.

[128] Kollman, G.E., J.G. Streeter, D.L. Jeffers and R.B. Curry, 1979. Accumulation and distribution of mineral nutrients, carbohydrate, and dry matter in soybean plants as influenced by reproductive sink size. Crop Sci 19: 729–734.

[129] Kremer, R. J. 1993. Management of weed seed banks with microorganisms. Ecol. Appl. 3:42–52.

[130] Kremer, R.J., Begonia, M.F.T., Stanley, L., Lanham, E.T., 1990. Characterization of rhizobacteria associated with weed seedlings. Appl. Environ. Microbiol. 56, 1649–1655.

[131] Kristensen L., Olsen J. and Weiner J. 2008. Crop density, sowing pattern, and nitrogen fertilization effects on weed suppression and yield in spring wheat. Weed Sci. 56:97-102

[132] Kropff, M.J., and H.H. van Laar. 1993. Modeling crop-weed interactions. CAB International, Wallingford, UK.

[133] Kruidhof, H. M., Bastiaans L. and Kropff M. J. 2008. Ecological weed management by cover cropping, effects on weed growth in autumn and weed establishment in spring. Weed Res., 48, 492-502.

[134] Kruse, M., Strandberg, M. and Strandberg, B. 2000. Ecological effects of allelopathic plants – A Review. National Environmental Research Institute, Silkeborg, Denmark. 66 pp. – NERI Technical Report No. 315.

[135] Kumar, V., D. C. Brainard, and R. R. Bellinder. 2008. Suppression of Powell amaranth (*Amaranthus powellii*), shepherd's-purse (*Capsella bursa-pastoris*), and corn chamomile (*Anthemis arvensis*) by buckwheat residues: role of nitrogen and fungal pathogens. Weed Sci. 56:271–280.

[136] Leather, G.R., 1983a. Sunflowers (*Helianthus annuus*) are allelopathic to weeds. Weed Sci. 31, 37–42.

[137] Leather, G.R., 1983b. Weed control using allelopathic crop plants. J. Chem. Ecol. 9, 983–990.

[138] Leather, G.R., 1987. Weed control using allelopathic sunflowers and herbicide. Plant Soil 98, 17–23.

[139] Lemerle, D., Verbeek, B., Cousens, R.D., Coombes, N.E., 1996. The potential for selecting wheat varieties strongly competitive against weeds. Weed Res. 36, 505–513.

[140] Li, Y., Sun, Z., Zhuang, X., Xu, L., Chen, S., Li, M., 2003. Research progress on microbial herbicides. Crop Prot. 22, 247–252.

[141] Liebman, M., and E. Gallandt. 1997. Many little hammers: ecological approaches for management of crop-weed interactions. p. 291–343. *In* L.E. Jackson (ed.) Ecology in Agriculture. Academic Press, San Diego.

[142] Lindquist, J. L. and D. A. Mortensen. 1998. Tolerance and velvetleaf (*Abutilon theophrasti*) suppressive ability of two old and two modern corn (*Zea mays*) hybrids. Weed Sci. 46:569–574.

[143] Lindquist, J. L. and M. J. Kropff. 1996. Applications of an ecophysiological model for irrigated rice (*Oryza sativa*)—*Echinochloa* competition. Weed Sci 44:52–56.

[144] Lindquist, J. L., D. A. Mortensen, and B. E. Johnson. 1998. Mechanisms of corn tolerance and velvetleaf suppressive ability. Agron. J. 90:787–792.

[145] Lovett, J. V. and Hoult, A. H. C. 1995. Allelopathy and self-defence in barley. In Allelopathy. Organisms, Processes, and Applications. (ed. Inderjit, K. M. M. Dakshini and F. A. E. (eds.).), pp. 170-183. American Chemical Society, Washington, DC.

[146] Lybecker D. W., Schweizer E. E. and King R. P. 1988. Economic analysis of four weed management systems. Weed Sci. 36, 846–849.

[147] Macias, F. A. 1995. Allelopathy in the search for natural herbicides models. In Allelopathy. Organisms, Processes and Applications (ed. K. M. M. Inderjit and E. F.A.), pp. 310-329. American Chemical Society.

[148] Macias, F. A., Simonet, A. M., Galindo, J. C. G., Pacheco, P. C. and Sanchez, J. A. 1998. Bioactive polar triterpenoids from Melilotus messanensis. Phytochemistry 49, 709-717.

[149] Macias, F.A., Varela, R.M., Torres, A., Molinillo, J.M.G., 1999. Potential of cultivar sunflowers (Helianthus annuus L.) as a source of natural herbicide template. In: Inderjit, Dakshini, K.M.M., Foy, C.L. (Eds.), Principles and Practices in Plant Ecology: Allelochemical Interactions. CRC Press, Boca Raton, FL, pp. 531–550.

[150] McDonald, G. K. 2003. Competitiveness against grass weeds in field pea genotypes. Weed Res. 43: 48–58.

[151] McWhorter, C.G., and E.E. Hartwig. 1972. Competition of johnsongrass and cocklebur with six soybean varieties. Weed Sci. 20: 56–59.

[152] Mesbah, A. O. and S. D. Miller. 1999. Fertilizer placement affects jointed goatgrass (Aegilops cylindrica) competition in winter wheat (Triticum aestivum). Weed Technol. 13: 374–377.

[153] Meschede, D. K., Ferreira, A. B. and Ribeiro Junior, C. C. 2007. Evaluation of weed suppression using different crop covers under Brazilian cerrado soil conditions. Planta Daninha, 25, 465-471.

[154] Miller, D. A. 1996. Allelopathy in forage crop systems. Agronomy Journal 88: 854-859.

[155] Mohammadi G. R. 2007. Growth parameters enhancing the competitive ability of corn (Zea mays L.) against weeds. Weed Biology and Management. 7: 232-236.

[156] Mohammadi G. R. and F. Amiri. 2011. Critical period of weed control in soybean (Glycine max) as influenced by starter fertilizer. Australian Journal of Crop Science. 5: 1350-1355.

[157] Mohammadi G. R. 2012. Living mulch as a tool to control weeds in agroecosystems: A Review. In: Price, A. J. (Ed.), Weed Control. InTech Press, Croatia. pp. 75-100.

[158] Mohammadi G. R., Kahrizi, D. and Rashidi, N. 2009. Evaluation of sorghum (Sorghum bicolor (L.) Moench) ability to control weeds in corn (Zea mays L.). Korean Journal of Weed Science. 29: 104-111.

[159] Mohammadi, G. R. 2009. The effects of legumes as living mulches on weed control and plant traits of corn (Zea mays L.). Korean Journal of Weed Science 29: 222-228.

[160] Mohammadi, G. R. 2010. Weed control in corn (Zea mays L.) by hairy vetch (Vicia villosa L.) interseeded at different rates and times. Weed Biology and Management 10: 25-32.

[161] Mohammadi, G. R., M. E. Ghobadi and S. Sheikheh Poor. 2012. Phosphate biofertiliz-
       er, row spacing and plant density effects on corn (*Zea mays* L.) yield and weed
       growth. American Journal of Plant Sciences. 3: 425-429.

[162] Mohammadi G. R., Mozafari S., Ghobadi M. E. and Najaphy A. 2012. Weed suppres-
       sive ability and productivity of berseem clover interseeded at different rates and
       times in corn field. The Philippine Agricultural Scientist 95:146-152.

[163] Mohan Babu R., Sajeena A., Seetharaman K., Vidhyasekaran P., Rangasamy P., Som
       Prakash M, Senthil Raja A. and Biji K. R. 2003. Advances in ioherbicides develop-
       ment—an overview. Crop Protection 22: 253–260.

[164] Moharramipour, S., Takeda, K., Sato, K., Yoshida, H. and Tsumuki, H. 1999. Inheri-
       tance of gramine in barley. Euphytica 106, 181-185.

[165] Mohler, C. L. 1996. Ecological bases for the cultural control of annual weeds. J. Prod.
       Agric. 9: 468–474.

[166] Mohler, C. L. 2001. Enhancing the competitive ability of crops. Pages 269–321 in M.
       Liebman, C. L. Mohler, and C. P. Staver, eds. Ecological Management of Agricultural
       Weeds. Cambridge, UK: Cambridge University Press.

[167] Mohler, C. L. and J. R. Teasdale. 1993. Response of weed emergence to rate of *Vicia
       villosa* Roth and *Secale cereale* L. residue. Weed Res. 33:487–499.

[168] Mohler, C. L. and M. B. Callaway. 1991. Effects of tillage and mulch on emergence
       and survival of weeds in sweet corn. J. Appl. Ecol. 29:21–34.

[169] Monks, D.W., and L.R. Oliver. 1988. Interactions between soybean (*Glycine max*) cul-
       tivars and selected weeds. Weed Sci. 36:770–774.

[170] Morris, M.J., 1997. Impact of the gall-forming rust fungus *Uromycladium tepperianum*
       on the invasive tree *Acacia saligna* in South Africa. Biol. Control 10, 75-82.

[171] Mutsaers, H.J.W. 1980. The effect of row orientation, date and latitude on light ab-
       sorption by row crops. J. Agric. Sci. 95:381–386.

[172] Narwal, S. S. 1996. Potentials and prospects of allelopathy mediated weed control for
       sustainable agriculture. In Allelopathy in Pest Management for Sustainable Agricul-
       ture. Procceding of the International Conference on Allelopathy, vol. ll (ed. S. S. Nar-
       wal and P. Tauro), pp. 23-65. Scientific Publishers, Jodhpur.

[173] Navas Marie-Laure and Moreau-Richard J. 2005. Can traits predict the competitive
       response of herbaceous Mediterranean species? Acta Oecologica. 27: 107-114.

[174] Niemeyer, H. M. and Jerez, J. M. 1997. Chromosomal location of genes for hydroxa-
       mic acids accumulation in *Triticum aestivum* L (wheat) using wheat aneuploids and
       wheat substitution lines. Heredity 79, 10-14.

[175] Nimbal, C.I., J.F. Pedersen, C.N. Yerkes, L.A.Weston, and S.C.Weller. 1996a. Phytotoxicity and distribution of sorgoleone in grain sorghum germplasm. J. Agric. Food Chem. 44:1343–1347.

[176] Nimbal, C.I., Yerkes, C.N., Weston, L.A., Weller, S.C., 1996b. Herbicidal activity and site of action of the natural product sorgoleone. Pest Chem. Physiol. 54, 73–83.

[177] O'Donovan, J. T. and Newman, J. C. 1996. Manipulation of canola (*Brassica rapa*) plant density and herbicide rate for economical and sustainable weed management. Proc. of the 2nd International Weed Control Congress, 25–28 June 1996, Copenhagen, Denmark. Department of Weed Control and Pesticide Ecology, Flakkebjerg, DK-4200 Slagelse, Denmark. pp. 969–974.

[178] O'Donovan, J. T., Harker, K. N., Clayton, G. W. and Hall, L. M. 2000. Wild oat (*Avena fatua*) interference in barley (*Hordeum vulgare*) is influenced by barley variety and seeding rate. Weed Technol. 14: 624–629.

[179] Oehrens, E., 1977. Biological control of the blackberry through the introduction of rust, *Phragmidium violaceum*, in Chile. FAO Plant Protect. Bull. 25, 26-28.

[180] Olofsdotter, M., L.B. Jensen, and B. Courtois. 2002. Improving crop competitive ability using allelopathy—an example from rice. Plant Breed. 121:1–9.

[181] Olofsdotter, M., Navarez, D. and Rebulanan, M. 1997. Rice allelopathy - Where are we and how far can we get? In The 1997 Brighton Crop protection Conference, vol. l, pp. 99-104, Brighton.

[182] Olsen, J. and J. Weiner. 2007. The influence of *Triticum aestivum* density, sowing pattern and nitrogen fertilization on leaf area index and its spatial variation. Basic Appl. Ecol. 8:252–257.

[183] Ominski, P.D., M.H. Entz, and N. Kenkel. 1999. Weed suppression by *Medicago sativa* in subsequent cereal crops: A comparative survey. Weed Sci. 47:282–290.

[184] Openshaw, S.J., H.H. Hadley and C.E. Brokoski, 1979. Effects of pod removal upon seeds of nodulations and non-nodulating soybean lines. Crop Sci 19: 289–290.

[185] Overland, L. 1966. The role of allelopathic substances in the "smother crop" barley. Amer. J. Bot. 53, 423-432.

[186] Paolini, R., M. Principi, R. J. Froud-Williams, S. Del Puglia, and E. Binacardi. 1999. Competition between sugarbeet and *Sinapis arvensis* and *Chenopodium album*, as affected by timing of nitrogen fertilization. Weed Res. 39:425–440.

[187] Petersen, J. 2003. Weed-spring barley competition for applied nitrogen in pig slurry. Weed Res. 43:33–39.

[188] Petho, M. 1992. Occurrence and physiological role of benzoxazinones and their derivates. lV. isolation of hydroxamic acids from wheat and rye root secretions. Acta Agronomica Hungarica 41, 167-175.

[189] Phatak, S. C. 1992. An integrated sustainable vegetable production system. Hort. Sci. 27, 738–741.

[190] Qasem, J. R. 1992. Root growth, development and nutrient uptake of tomato (*Lycopersicon esculentum*) and *Chenopodium album*. Weed Res. 33:35–42.

[191] Rajalahti, R. M., Bellinder, R. R. and Hoffman, M. P. 1999. Time of hilling and interseeding affects weed control and potato yield. Weed Science 47, 215-225.

[192] Rajcan, I., M. AghaAlikhani, C. J. Swanton, and M. Tollenaar. 2002. Development of redroot pigweed is influenced by light spectral quality and quantity. Crop Sci. 42:1930–1936.

[193] Randall, H.W.,Worsham, D., Blum,W., 1989. Allelopathic potential of legume debris and aqueous extracts. Weed Sci. 37, 674–679.

[194] Rasmussen, K. 2002. Influence of liquid manure application method on weed control in spring cereals. Weed Res. 42:287–298.

[195] Reich, P.B., M.B. Walters and D.S. Ellsworth, 1997. From tropics to tundra: Global convergence in plant functioning. Proc Natl Acad Sci (USA) 94: 13730–13734.

[196] Rice, E.L. 1984. Allelopathy. 2nd ed. Academic Press, New York.

[197] Rice, E.L. 1995. Biological control of weeds and plant diseases. Univ. Oklahoma Press, Norman.

[198] Rimando, A.M., Olofsdotter, M., Dayan, F.E., Duke, S.O., 2001. Searching for rice allelochemicals: an example of bioassay-guided isolation. Agron. J. 93, 16–20.

[199] Rizvi, S. H. and V. Rizvi. 1992. Allelopathy—Basic and Applied Aspects. 1st ed. London: Chapman and Hall. 480 p.

[200] Rodenburg, J., Saito, K., Kakai, R.G., Toure, A., Mariko, M., Kiepe, P., 2009. Weed competitiveness of the lowland rice varieties of NERICA in the southern Guinea Savanna. Field Crops Res. 114, 411–418.

[201] Ross, M.A. and J.L. Harper, 1972. Occupation of biological space during seedling establishment. J Ecol 60: 77–88.

[202] Ross, S. M., King, J. R., Izaurralde, R. C. and O'Donovan, J. T. 2001. Weed suppression by seven clover species. Agronomy Journal 93, 820-827.

[203] Ross, M.A. and Lembi, C.A., 1985. Applied Weed Science. Burgees Pub. Co.

[204] Saito K., Azoma K. and Rodenburg J. 2010. Plant characteristics associated with weed competitiveness of rice under upland and lowland conditions in West Africa. Field Crops Res., 116: 308-317.

[205] Samson, R. A. 1991. The weed suppressing effects of cover crops. Pages 11–22 in Fifth Annual REAP Conference, Macdonald College, Ste.-Anne-de-Bellevue, Quebec.

[206] Sankula, S., M. J. VanGessel, and R. R. Mulford. 2004. Corn leaf architecture as a tool for weed management in two production systems. Weed Sci. 52:1026–1033.

[207] Santos, B. M., J. A. Dusky, W. M. Stall, D. G. Shilling, and T. A. Bewick. 1997. Influence of smooth pigweed and common purslane densities on lettuce yields as affected by phosphorus fertility. Proc. Fla. State Hortic. Soc. 110:315–317.

[208] Sardi, K. and I. Beres. 1996. Effects of fertilizer salts on the germination of corn, winter wheat, and their common weed species. Commun. Soil Sci. Plant Anal. 27:1227–1235.

[209] Sattin, M., M. C. Zuin, and I. Sartorato. 1994. Light quality beneath field grown maize, soybean, and wheat canopies—red:far red variations. Physiol. Plant. 91:322–328.

[210] Schreiber, M.M. 1992. Influence of tillage, crop rotation, and weed management on giant foxtail (*Setaria faberi*) population dynamics and corn yield. Weed Sci. 40:645–653.

[211] Schwinning, S. and J. Weiner. 1998. Mechanisms determining the degree of size asymmetry in competition among plants. Oecologia 113:447–455.

[212] Se'ne, M., T. Dore', and C. Gallet. 2001. Relationship between biomass and phenolic production in grain sorghum grown under different conditions. Agron. J. 93:49–54.

[213] Sexsmith, J. J. and U. J. Pittman. 1963. Effect of nitrogen fertilizers on germination and stand of wild oats. Weeds 11:99–101.

[214] Sharma, V. and N. N. Angiras. 1996a. Effect of row orientations, row spacings and weed-control methods on light interception, canopy temperature and productivity of wheat (*Triticum aestivum*). Indian J. Agron. 41:390–396.

[215] Sharma, V. and N. N. Angiras. 1996b. Light interception, weed growth and productivity of irrigated wheat as influenced by crop geometry and weed control methods. Indian J. Plant Physiol. 1:157–162.

[216] Sheaffer, C. C., Gunsolus, J. L., Grimsbo Jewett, J. and Lee, S. H. 2002. Annual *Medicago* as a smother crop in soybean. Journal of Agronomy and Crop Science 188, 408-416.

[217] Shrefler, J. W., J. A. Dusky, D. G. Shilling, B. J. Brecke, and C. A. Sanchez. 1994. Effects of phosphorus fertility on competition between lettuce (*Lactuca sativa*) and spiny amaranth (*Amaranthus spinosus*). Weed Sci. 42:556–560.

[218] Shrestha, A. and M. Fidelibus. 2005. Grapevine row orientation affects light environment, growth, and development of black nightshade (*Solanum nigrum*). Weed Sci. 53: 802–812.

[219] Sibuga, K. P. and J. D. Bandeen. 1980. Effects of various densities of green foxtail (*Setaria viridis* L. Beav.) and lambsquarters (*Chenopodium album*) on N uptake and yields of corn. E. Afric. Agric. For. J. 45:214–221.

[220] Singh H. P., Batish D. R. and Kohli R.K. 2003. Allelopathic interactions and allelo-
       chemicals: New possibilities for sustainable weed management. Critical Reviews in
       Plant Sciences. 22: 239-311.

[221] Singh, H.P., D.R. Batish, S. Kaur, N. Setia, and R.K. Kohli. 2005. Effects of 2-benzoxa-
       zolinone on the germination, early growth and morphogenetic response of mung
       bean (*Phaseolus aureus*). Ann. Appl. Biol. 147:267–274.

[222] Sinoquet, H., Caldwell, R.M. 1995. Estimation of light capture and partitioning in in-
       tercropping systems. P.79-80. In H. Sinoquet, P. Cruz (Eds.) Ecophysiology of tropi-
       cal intercropping. Inst. Natl. de la Recherche Agronomique (INRA), Paris.

[223] So, Y. F., M. M. Williams, II, J. K. Pataky, and A. S. Davis. 2009. Principal canopy fac-
       tors of sweet corn and relationships to competitive ability with wildproso millet
       (*Panicum miliaceum*). Weed Sci. 57:296–303.

[224] Spies, J.M., T.D. Warkentin and S.J. Shirtliffe. 2011. Variation in field pea (*Pisum sati-
       vum*) cultivars for basal branching and weed competition. Weed Sci. 59:218-223.

[225] Steinbauer, G. P. and B. Grigsby. 1957. Interaction of temperature, light, and moisten-
       ing agent in the germination of weed seeds. Weeds 5:175–182.

[226] Steinmaus, S., Elmore, C. L., Smith, R. J., Donaldson, D., Weber, E. A., Roncoroni, J.
       A. and Teasdale, J. R. 1993. Reduced-herbicide weed management systems for no-till-
       age corn (*Zea mays*) in a hairy vetch (*Vicia villosa*) cover crop. Weed Technol. 7, 879–
       883.

[227] Stephen, O.D., Lydon, J., 1987. Herbicides from natural compounds. Weed Technol. 1
       (2), 122–128.

[228] Supasilapa, S., Steer, B.T., Milroy, S.P., 1992. Competition between lupin (*Lupinus an-
       gustifolia* L.) and great brome (*Bromus diandrus* Roth.): development of leaf area, light
       interception and yields. Aust. J. Exp. Agric. 32, 71–81.

[229] Sweeney A. E., Renner K. A., Laboski C. and Davis A. 2008. Effect of fertilizer nitro-
       gen on weed emergence and growth. Weed Sci 56:714-721.

[230] Teasdale, J. R. 1995. Influence of narrow row/high population corn (*Zea mays*) on
       weed control and light transmission. Weed Technol. 9:113–118.

[231] Teasdale, J. R. 1996. Contribution of cover crops to weed management in sustainable
       agricultural systems. J. Prod. Agric. 9, 475–479.

[232] Teasdale, J.R., 1998. Influence of corn (*Zea mays*) population and row spacing on corn
       and velvetleaf (*Abutilon theophrasti*) yield. Weed Sci. 46, 447–453.

[233] Teasdale, J. R. 2003. Principles and practices of using cover crops in weed manage-
       ment systems. In, R. Labrada (ed), Weed Management for Developing Countries.
       Food and Agriculture Organization of the United Nations (FAO), Rome, Italy. pp.
       169-178.

[234] Teasdale, J. R. and C. L. Mohler. 1993. Light transmittance, soil temperature, and soil moisture under residue of hairy vetch and rye. Agron. J. 85, 673-680.

[235] Teyker, R. H., H. D. Hoelzer, and R. A. Liebl. 1991. Maize and pigweed response to nitrogen supply and form. Plant Soil 135:287-292.

[236] Tollenaar M., Nissanka S.P., Aguilera A.,Weise S.F. and Swanton C.J. 1994. Effect of weed interference and soil nitrogen on four maize hybrids. *Agron. J.* 86, 596–601.

[237] Townley-Smith, L. and Wright, A. T. 1994. Field pea cultivar and weed response to crop seed rate in western Canada. Can. J. Plant Sci. 74: 387–393.

[238] Trujillo, E.E., 1985. Biological control of Hamakua pa-makani with *Cercosporella* sp. in Hawaii. In: Delfosse, E.S. (Ed.), Proceedings of the VI International Symposium on Biological Control of Weeds. Agriculture Canada, Ottawa, pp. 661-671.

[239] Tuncw, I. and S. Sahinkaya. 1998. Sensitivity of two greenhouse pests to vapours essential oils. Entomol. Exp. Appl. 86:183–187.

[240] Tworkoski, T., 2002. Herbicide effects of essential oils. Weed Science, 50: 425-431.

[241] Van Acker R. C., Swanton C. J. and Weise S. F. 1993. The critical period of weed control in soybeans [*Glycine max* (L.) Merr.]. Weed Sci 41: 194-200.

[242] Van Delden, A., L. A. Lotz, L. Bastiaans, A. C. Franke, H. G. Smid, R.M.W. Groeneveld, and M. J. Kropff. 2002. The influence of nitrogen supply on the ability of wheat and potato to suppress Stellaria media growth and reproduction. Weed Res. 42:429–445.

[243] Vandermeer, J., 1989. The Ecology of Intercropping. Cambridge University Press.

[244] van der Werf, A., M.V. Nuenen, A.J. Visser, and H. Lambers. 1993. Contribution of physiological and morphological plant traits to a species' competitive ability at high and low nitrogen supply. Oecologia 94:434–440.

[245] Vera C.L., Woods S.M. and Raney J.P. 2006. Seeding rate and row spacing effect on weed competition, yield and quality of hemp in the Parkland region of Saskatchewan. Can. J. Plant Sci. 86: 911-915.

[246] Verma, R., H. R. Agarwal, and V. Nepalia. 1999. Effect of weed control and phosphorus on crop-weed competition in fenugreek (*Trigonella foenumgraecum*). Indian J. Weed Sci. 31: 265–266.

[247] Vokou, D. 2007. Allelochemicals, allelopathy and essential oils: a field in search of definitions and structure. Allelop. J. 19: 119–134.

[248] Vrabel, T. E. 1983. Effects.of suppressed white clover on sweet corn yield and nitrogen availability in a living mulch cropping system. Ph.D Thesis (Diss. abstr. DA 8321911), Cornell Univ., Ithaca, NY.

[249] Walker, R. H. and G. A. Buchanan. 1982. Crop manipulation in integrated weed management systems. Weed Sci. 30 (Suppl. 1): 17–24.

[250] Wall, D. A. and L. Townley-Smith. 1996. Wild Mustard (*Sinapsis arvensis*) responses to field pea (*Pisum sativum*) cultivar and seeding rate. Can. J. Plant Sci. 76: 907–914.

[251] Wallace, D.H., K.S. Yourstone, P.N. Masaya and R.W. Zobel. 1993. Photoperiod gene control over partitioning between reproductive vs. vegetative growth. Theor Appl Genet 86: 6–16

[252] Wang G, Ehlers JD, Marchi ECS *and* McGiffen ME, Jr. 2006. Competitive ability of cowpea (*Vigna unguiculata*) genotypes with different growth habit. Weed Science 54: 775–782.

[253] Watson, A.K., 1991. The classical approach with plant pathogens. In: TeBeest, D.O. (Ed.), Microbial Control of Weeds. Chapman and Hall, New York, pp. 3-23.

[254] Watson, P. R., Derksen, D. A., Van Acker, R. C., and Blrvine, M. C., 2002. The contribution of seed, seedling, and mature plant traits to barley cultivar competitiveness against weeds. Proceedings of the 2002 National Meeting – Canadian Weed Science Society. 49-57.

[255] Watson, P.R., D.A. Derksen, and R.C. Van Acker. 2006. The ability of 29 barley cultivars to compete and withstand competition. Weed Sci. 54:783–792.

[256] Weaver S. E., Kropff M. J. and Groeneveld R. M. W. 1992. Use of ecophysiological models for crop-weed interference: the critical period of weed interference. Weed Sci 40: 302-307.

[257] Weiner, J. 1990. Asymmetric competition in plant populations. Trends Ecol. Evol. 5:360–364.

[258] Weiner, J., 1986. How competition for light and nutrients affects size variability in ipomoea-tricolor populations. Ecology 67, 1425–1427.

[259] Weiner, J., H. W. Griepentrog, and L. Kristensen. 2001. Suppression of weeds by spring wheat (*Triticum aestivum*) increases with crop density and spatial uniformity. J. Appl. Ecol. 38:784–790.

[260] Welles, J.M., Norman, J.M., 1991. Instrument for indirect measurement of canopy architecture. Agron. J. 83, 818–825.

[261] Weston, L. A. 1996. Utilization of allelopathy for weed management in agroecosystems. Agron. J. 88:860–866.

[262] Weston, L. A. 2005. History and current trends in the use of allelopathy for weed management. Hort. Technology 15, 529–534.

[263] Weston, L.A., R. Harmon, and S. Mueller. 1989. Allelopathic potential of sorghum–sudangrass hybrid (sudex). J. Chem. Ecol. 15:1855–1865.

[264] Westra, E. P. 2010. Can Allelopathy be incorporated into agriculture for weed suppression? http,//www.colostate.edu/Depts/Entomology/courses/en570/papers_2010/westra.pdf.

[265] White R. H., Worsham A. D. and Blum U. 1989. Allelopathic potential of legume debris and aqeous extracts. Weed Sci. 37, 674–679.

[266] Williams, W.A., Loomis, R.S., Duncan, W.G., Dovert, A., Nunez, F., 1968. Canopy architecture at various population densities and the growth of grain of corn. Crop Sci. 8: 303–308.

[267] Wortmann, C.S. 1993. Contribution of bean morphological characteristics to weed suppression. Agron. J. 85:840–843.

[268] Wu, H., J. Pratley, D. Lemerle, and T. Haig. 2000. Evaluation of seedling allelopathy in 453 wheat (*Triticum aestivum*) accessions against annual ryegrass (*Lolium rigidum*) by equal-compartment-agar method. Aust. J. Agric. Res. 51:937–944.

[269] Wu, H., J. Pratley,W. Ma, and T. Haig. 2003. Quantitative trait loci and molecular markers associated with wheat allelopathy. Theor. Appl. Genet. 107:1477–1481.

[270] Wu, H., pratley, J., Lemerle, D. and Haig, T. 1999. Crop cultivars with allelopathic capability. Weed Research. 39: 171-180.

[271] Wu, H., Pratley, J., Lemerle, D., Haig, T. and Verbeek, B. 1998. Differential allelopathic potential among wheat accessions to annual ryegrass. In. Proceedings 9th Australian Agronomy Conference, Wagga Wagga, Australia, 567-571 .

[272] Yenish, J.P., J.C. Doll, and D.G. Buhler. 1992. Effects of tillage on vertical distribution and viability of weed seed in soil. Weed Sci. 40:429–433.

[273] Yoshida, H., Iida, T., Sato, K., Moharramipour, S. and Tsumuki, H. 1997. Mapping a gene for gramine synthesis. Barley genetics newsletter 27, 22-24.

[274] Yoshida, H., Tsumuki, H., Kanehisa, K. and Corucuera, L. J. 1993. Release of gramine from the surface of barley leaves. Phytochemistry 34, 1011-1013.

[275] Zhao, D.L., Atlin, G.N., Bastiaans, L., Spiertz, J.H.J., 2006. Cultivar weed-competitiveness in aerobic rice: heritability, correlated traits, and the potential for indirect selection in weed-free environments. Crop Sci. 46: 372–380.

Chapter 6

Integrated Pest Management

C.O. Ehi-Eromosele, O.C. Nwinyi and O.O. Ajani

Additional information is available at the end of the chapter

1. Introduction

Integrated Pest Management (IPM) has had a varied history, with different definitions. It has been implemented under an array of different connotations (Lewis *et al.*, 1997). The term was earlier used as "integrated control" by Bartlett (Bartlett, 1956) and was further elaborated on by Stern and co-workers (Stern *et al.*, 1959). In reference to the concept of integrating the use of biological and other controls in complementary ways, the term was later broadened to embrace coordinated use of all biological, cultural, and artificial practices (van den Bosch and Stern, 1962). The term "IPM," under various authors have advocated for the principle of incorporating, the full array of pest management practices with production objectives in a total systemic approach. Nonetheless, there is no universally agreed definition of IPM.

Alastair (2003) conceived IPM as a method of rationalizing pesticide use to prevent or delay the resurgence of pest populations, that had become resistant to pesticides, and to protect beneficial insects. Today, concerns about pesticide residues in the food chain and in the environment have led to alternative definitions that exclude the use of conventional pesticides. Nevertheless, there are broad agreements on the core principles of IPM. These include IPM being:

- An integrated scheme due to the methods of control that are seen as component technologies rather than alternatives.

- emphasizes pest management, within a balanced system whereas control strategy suggests direct intervention with little concern for sustainability (Alastair, 2003).

In principle, IPM may be defined as a flexible and holistic system. This views the agro-ecosystem as an interrelated whole, that utilizes a variety of biological, cultural, genetic, physical, and chemical techniques that hold pests below economically damaging levels with a minimum disruption to the cropping ecosystem and the surrounding environment (Malena, 1994). In the definition of FAO (2012), integrated pest management (IPM) is an ecosystem approach to crop production and protection that combines different management strategies and practices to grow healthy crops and minimize the use of pesticides. IPM therefore, utilizes the best mix of control tactics for a given pest problem when compared with the crop yield, profit and safety of other alternatives (Kenmore *et al.*, 1985). Other definitions of IPM according to the United States Environment Protection Agency (2012), involves an effective and environmentally sensitive approach to pest management that relies on a combination of common-sense practices. IPM could be a broad based ecological approach to structural and agricultural pest control that integrates pesticides/herbicides into a management system incorporating a range of practices for economic control of a pest. In addition, IPM, attempts to prevent infestation, to observation of patterns of infestation when they occur, and to intervene (without poisons) when one deems necessary. Finally, IPM is the intelligent selection and use of pest control actions that ensures favourable economic, ecological and sociological consequences (Sandler, 2010).

1.1. Succinct history of IPM

IPM is not a new philosophy. The concept has been around since the 1920's when cotton pest management program was developed. Under this scheme, insect control was "supervised" by qualified entomologists, and insecticide applications were based on conclusions reached from periodic monitoring of pest and natural-enemy populations. This was viewed as an alternative to calendar-based insecticide programs. Supervised control is based on a sound knowledge of the ecology and analysis of the projected trends in pest and natural-enemy populations. In supervised control, (integrated control) the best mix of chemical and biological controls is sought and identified for a given insect pest. The chemical insecticides are used in a manner that is least disruptive to biological control. The chemical controls are applied only after regular monitoring indicates that a pest population had reached an economic threshold level. Thus, such treatment is required to prevent the population from reaching an economic injury level where economic losses would exceed the cost of the artificial control measures.

Typically, the main aim of IPM programmes is on agricultural insect pests (IPM Guidelines, 2009). Although, originally developed for agricultural pest management, IPM programmes are now developed to encompass diseases, weeds, and other pests that may interfere with the management objectives of sites such as residential and commercial structures, lawn and turf areas, and home and community gardens. IPM programs use current, comprehensive information on the life cycles of pests and their interaction with the environment. This information, in combination with available pest control methods, is used to manage pest damage by the most economical means, and with the least possible hazard to people, property, and the environment.

The IPM approach can be applied to both agricultural and non-agricultural settings, such as the home, garden, and workplace. IPM takes advantage of all appropriate pest management options including, the judicious use of pesticides.

2. IPM – A philosophy

2.1. A pest management strategy

Integrated Pest Management (IPM) is a philosophy that involves the management of a pest instead of controlling or eradicating a pest. It requires a greater knowledge of the pest, crop and the environment. Therefore, its strategy focuses on harnessing inherent strengths within ecosystems and directing the pest populations into acceptable bounds rather than toward eliminating them. This strategy avoids undesirable short term and long term ripple effects and will ensure a sustainable future (Lewis et al., 1997).

IPM programs should be operated with "pest management objectives" rather than "pesticide management objectives". Integrated pest management is a comprehensive long term pest management program based on knowledge of an ecosystem that weighs economic, environmental, and social consequences of interventions (Flint and van den Bosch, 1981). The foundation for pest management in agricultural systems should be an understanding and shoring up of the full composite of inherent plant defenses, plant mixtures, soil, natural enemies, and other components of the system. These natural "built in" regulators are linked in a web of feedback loops that are renewable and sustainable. The use of pesticides and other "treat-the-symptoms" approaches are unsustainable and should be the last option rather than the first line of defense. A pest management strategy should always start with the question "Why is the pest a pest?". It should also seek to address underlying weaknesses in ecosystems and/or agronomic practice(s) that have allowed organisms to reach pest status (Lewis et al., 1997).

2.2. An integrated process

Integration or compatibility among pest management tactics is central to Integrated Pest Management. Simply mixing different management tactics does not constitute IPM. Mixing the tactics arbitrarily may actually aggravate pest problems or produce other unintended effects. IPM recognizes there is no "cure-all" in pest control (dependence on any one pest management method will have undesirable effects). Reliance on a single tactic will favor pests that are resistant to that practice. In IPM, integrated control seeks to identify the best mix of chemical and biological controls for a given insect pest. The term "integrated" is thus synonymous with "compatibility."

2.3. Understanding pest biology and ecology

The determination of the correct cause of pest problem (understanding pest biology) and ecology is essential in manipulating the environment to the crop's advantage and to the detriment of the pest.

2.4. Acceptable pest levels

IPM recognizes that eradication of a pest is seldom necessary or even desirable, and generally not possible. The primary objective in pest management is not to eliminate a pest organism but to bring it into acceptable bounds (Lawal *et al.*, 1997). The emphasis is on control, not eradication. IPM holds that wiping out an entire pest population is often impossible, and the attempt can be expensive and environmentally unsafe. IPM programmes initial task is to establish acceptable pest levels, called action thresholds, and apply controls where the thresholds are crossed. These thresholds are pest and site specific, meaning that it may be acceptable at one site to have for instance a weed such as white clover, but at another site it may not be acceptable. By allowing a pest population to survive at a reasonable threshold, selection pressure is reduced. This stops the pest gaining resistance to chemicals produced by the plant or applied to the crops. If many of the pests are killed, then any that have resistance to the chemical will form the genetic basis of the future, more resistant, population. By not killing all the pests there are some un-resistant pests left that will dilute any resistant genes that appear (Wikipedia, 2011).

2.5. IPM a continuum, not an end

Agriculture is a dynamic system that continually changes to changing crop production practices. IPM must continually change to meet pest management challenges. IPM is a continuum that will change with time. Every farmer practices some type of IPM, as long as they make progress to better its management. As new pest control techniques are discovered, the producer and crop advisor must adapt their pest control program to reflect these changes. What is considered a good IPM program today may be considered a chemical intensive program in a few years. Additionally, some good advice to the producer and crop advisor is to try the new changes on a limited scale, while becoming comfortable with the suggested practices before wide-scale changes are made.

3. IPM process

IPM is applicable to all types of agriculture and sites such as residential and commercial structures, lawn and turf areas, and home and community gardens. The process includes:

3.1. Proper identification pest damage and responsible pests

Identification must be the first objective. When the identity of a pest is not known, then, a strategy built to control the pest cannot be transferred from one site to another, primarily, because the pest species or strain (biotype) might behave differently. Thus, a solid foundation must be built on systematic, taxonomy, etiology, and spatial distribution (Irwin, 1999). Cases of mistaken identity may result in ineffective actions. If plant damage is due to over-watering, it could be mistaken for fungal infection, since many fungal and viral infections arise under moist conditions. This could lead to spray costs, but the plant would be no better off.

3.2. Pest and host life cycles biology

Understanding crop growth and development is an underlying principle of IPM. We cannot just focus on the pest. The interactions between crop and pest (as well as the environment)

are very important. To deplore an efficient IPM programme, literature and other data sources about the pest, the pest's life cycle, host range, distribution, movement, and basic biology will have to be researched. At the time you see a pest, it may be too late to do much
about it except maybe spray with a pesticide (Metcalf and Luckmann, 1994). Often, there is
another stage of the life cycle that is susceptible to preventative actions. For example, weeds
reproducing from last year's seed can be prevented with mulches and pre-emergent herbicide. Also, learning what a pest needs to survive allows you to remove these.

3.3. Monitor or sample the environment for pest populations

After the pest has been correctly identified, monitoring must begin before it becomes a problem. Sampling and monitoring methodologies must be designed and tested to provide the
ability for assessing instantaneous and dynamic aspects of the pest's density, activity, or incidence (Irwin, 1999). Understanding how the environment affects pest and crop development is very important. Understanding interactions with the environment allows crop
advisors to react to changing conditions. Environmental influences like drought stress influences pest management recommendations. When a crop is under stress it can be less capable
of dealing with the stress caused by insects that extract plant sap (e.g. aphids, leafhoppers)
and this stress may slightly lower the economic threshold. Weed populations which would
not normally cause an economic loss may do so under drought conditions when they compete with the crop for limited water.

The weather is notorious for affecting pest development and survival. Certain weather patterns may affect weed seed germination and explain why certain weeds are more abundant
during wet fall or springs.

3.4. Establish action threshold (economic, health, and aesthetic)

The question here is: how many are too many or how much can be tolerated? In some cases,
there are standardized numbers of pests that can be tolerated. Soybeans are quite tolerant of
defoliation, so if there are a few caterpillars in the field and their population may not be increasing dramatically; thus, no urgent action may be necessary. Conversely, there is a point
at which an action must be taken to control cost. For instance the farmer can control cost at
the point when the cost of damage by the pest is more than the cost of control. This is an
economic threshold. Tolerance of pests varies according to the health hazard (low tolerance)
or merely a cosmetic damage (high tolerance in a non-commercial situation). Different sites
may also have varying requirements based on specific areas. For instance, white clover may
be perfectly acceptable on the sides of a tee box on a golf course, but unacceptable in the
fairway where it could cause confusion in the field of play (Purdue University, 2006).

3.5. Choose an appropriate combination of management tactics

The word 'integrated' in IPM initially referred to the simultaneous use or integration of any
number of tactics in combination, with focus on maintaining a single pest species below its
economic injury level. Although, in theory a single strategy results from the simultaneous

integration of several tactics, in practice, the integration actually occurs in a step-wise, time-delayed fashion. Several of the tactics are compatible, but some are not. Certainly the tactics of biological control, habitat manipulation, and legal control go alongside. The tactic of host resistance can stand alone or be combined with the other tactics just mentioned. Chemical control is generally compatible with host resistance. Thus, a management strategy integrates one or several compatible tactics into a single package (Irwin, 1999).

3.6. Evaluate and record results

Evaluation is often one of the most important steps in Integrated Pest Management (Bennett *et al.*, 2005). It is the process of reviewing an IPM program and the results it has generated. Asking the following questions is useful: Did the steps one took effectively control the population? Was this method safe enough? Where there any expected side effects? What is the next step? Understanding the effectiveness of the IPM program allows the site manager to make modifications to the IPM plan prior to pests reaching the action threshold and requiring action again.

4. Pest management tactics

There are different pest management tactics to suppress pests. They include host resistance, chemical, biological, cultural, mechanical, sanitary and mechanical controls. The primary pest management tactic involves maximization of built-in pest reduction features of an ecosystem. Molecular or genetic mechanisms are potentially manifested in a number of these more specific tactics. Each category, discussed below, employs a different set of mechanisms for suppressing populations.

4.1. Chemical control

The therapeutic approach of killing pest organisms with toxic chemicals has been the prevailing pest control strategy for over 50 years. Safety problems and ecological disruptions continue to ensue (Wright, 1996), and there are renewed appeals for effective, safe, and economically acceptable alternatives (Benbrook, 1996). Synthetic chemical pesticides are the most widely used method of pest control. The four major problems encountered with conventional pesticides are toxic residues, pest resistance, secondary pests, and pest resurgence (Lewis, 1997). The use of natural pesticides and organophosphates that are more environmentally friendly are encouraged and synthetic pesticides should only be used as a last resort or only used as required and often only at specific times in a pests life cycle.

4.2. Biological control

This involves the use of other living things that are enemies of a pest in order to control it. Sometimes, the term "biological control" has been used in a broad context to encompass a full spectrum of biological organisms and biologically based products including pheromones, resistant plant varieties, and autocidal techniques such as sterile insects. IPM is mainly aimed at developing systems based on biological and non-chemical methods as much as possible.

4.3. Host plant resistance

This involves breeding varieties with desirable economic traits, but less attractive for pests, for egg laying and subsequent development of insect, disease or nematode. It also involves withstanding the infestation/infection or the reduction of pests to level that they are not large numbers during the plant growth period (Sharma, 2007).

4.4. Cultural measures

This involves practices that suppress pest problems by minimizing the conditions that favour their existence (water, shelter, food). Some of these factors are intrinsic to crop production while making the environment less favourable for survival, growth and reproduction of pest species. If followed in an appropriate manner, the cultural practices can provide significant relief from pests. The selection of appropriate site for the cultivation of field crops and fruit trees can reduce future infestation from insect pests. The culture should be selected in such a manner that it should be suitable for growing in the area and tolerant to important pests diseases of the area.

4.5. Mechanical control

This is the use of machinery and other tools to control pests. It involves agricultural practices like tillage, slash and burn, and hand weeding. The pruning of infested parts of fruits and forest trees and defoliation in certain crops help reduce the pest population. Chaffing of sorghum/maize stalks and burning of stubbles kills maize borer.

4.6. Sanitary control

Preventive practices are important part of an IPM programme. These include cleaning field equipment (i.e., tillage equipment, haying equipment, etc.), planting certified seeds and quarantine of infested crops or farmlands. These are methods used to prevent the introduction of a pest into the field.

4.7. Natural control

Natural control involves the enhancement of naturally occurring pest management methods to combat pests like using beneficial insects and diseases. Here, insecticides will only be used when they are economically feasible and it is apparent that natural enemies will not control the pests.

5. IPM: A multi-disciplinary approach

IPM is a management intensive philosophy which stresses a multidisciplinary approach. Pests interact with each other, the crop, and the environment. Similarly, pest and crop management disciplines must work together to develop control recommendations that reflect these interactions.

For example, management of the Soybean aphid includes entomologists who study the insect and their damage to soybean, agronomists that identify crop stage which are most vulnerable to soybean aphid damage, plant pathologists who study the viruses transmitted by aphid feeding, and soil scientists who study the aphids interaction with nutrient deficiencies.

6. Benefits of an IPM programme

The benefits of Integrated Pest Management are immense directly to farming and indirectly to society.

a. Integrated Pest Management (IPM) protects environment through elimination of unnecessary pesticide applications. In IPM, pesticides are used at the smallest effective dose when other methods of pest control have failed. Also, they are used in bringing a pest organism to acceptable bounds with as little ecological disruption as possible.

b. IPM improves profitability. Since IPM programme applies the most economical management pest tactics, profitability is ensured for the grower or farmer.

c. It reduces risk of crop loss by a pest. Applying pest management and monitoring tactics will also ensure the reduction of crop loss or damage by pests.

d. Long term sociological benefits of IPM would also emerge in areas of employment, public health, and well being of persons associated with agriculture.

7. Disadvantages of an IPM programme

In spite of the numerous benefits of IPM stated so far, there are also some drawbacks to it:

7.1. An IPM program requires a higher degree of management

Making the decision not to use pesticides on a routine or regular basis requires advanced planning and therefore, a higher degree of management. This planning includes attention to field histories to anticipate what the pest problems might be, selecting crop varieties which are resistant or tolerant to pest damage, choosing tillage systems that will suppress anticipated pest damage while giving the crop the greatest yield potential.

7.2. IPM can be more labor intensive

Consistent, timely and accurate field scouting takes time. However, it is this information that is necessary and is the corner stone of IPM programs. Without this information you cannot make intelligent management decision.

7.3. Success can be weather dependant

Weather can complicate IPM planning. For example you might want to lower herbicide rates and use row cultivation to manage weed pressure. However an extended wet period may reduce (or eliminate) the effectiveness of row cultivation. Therefore, good IPM planners will have a alternate plan for when these problems arise.

Author details

C.O. Ehi-Eromosele[1], O.C. Nwinyi[2*] and O.O. Ajani[1]

*Address all correspondence to: nwinyiobinna@gmail.com

1 Chemistry Department, School of Natural and Applied Sciences, College of Science and Technology, Covenant University, Canaan Land, Ota, Ogun State, Nigeria

2 Department of Biological Sciences, School of Natural and Applied Sciences, College of Science and Technology, Covenant University,, Canaan Land, Ota, Ogun State, Nigeria

References

[1] Alastair, O. (2003) Integrated Pest Management for Resource-Poor African Farmers: Is the Emperor Naked? World Development Vol. 31, No. 5, pp. 831–845

[2] Barlett, B. R. (1956) Natural predators. Can selective insecticides help to preserve biotic control Agric. Chem. 11, 42–44.

[3] Benbrook, C. M. (1996) Pest Management at the Crossroads (Consumer's Union, Yonkers, NY).

[4] Bennett, G.W., Owens, J.M., Corrigan, R.M. (2005) Truman's Scientific Guide to Pest Management Operations, 6th Edition, Purdue University, Questex, USA. pp. 10-12,

[5] Ehler, L. E.; Bottrell D. G. (2000) The illusion of integrated pest management. Issues in Science and Technology 16, 61-64.

[6] Flint, M. L. and van den Bosch, R. (1981) Introduction to Integrated Pest Management (Plenum, New York).

[7]   Food and Agricultural Organisation of the United Nations http://www.fao.org/agriculture/crops/core-themes/theme/pests/ipm/en/

[8]   IPM Guidelines . UMassAmherst: Integrated Pest Management, Agriculture and Landscape Program 2009 http://www.umass.edu/umext/ipm/publications/guidelines/index.html.

[9]   Irwin, M.E. (1999) Agricultural and Forest Meteorology 97: 235–248

[10] Kenmore, P.E., Heong, K.L., Putter, C.A (1985) Political, social and perceptual aspects of integrated pest management programmes. In:Lee, B.S., Loke, W.H., Heong, K.L. (Eds.), Integrated Pest Management in Asia. Malaysian Plant Protection Society, Kuala Lumpur, pp. 47- 66

[11] Lewis, W. J., van Lenteren, J. C., Sharad C. Phatak, and Tumlinson, J. H. (1997) Proc. Natl. Acad. Sci. USA 94: 12243–12248

[12] Malena, C. (1994) Gender issues in IPM in African agriculture. Socio-economic series no. 5. Chatham: Natural Resources Institute.

[13] Purdue University Turf Pest Management Correspondence Course, Introduction, 2006

[14] Robert L. M., William H. L. (1994). Introduction to Insect Pest Management. New York: John Wiley and Sons, Inc. pp. 266.

[15] Sandler, Hilary A. (2010). "Integrated Pest Management". Cranberry Station Best Management Practices 1 (1): 12–15.

[16] Stern, V. M., Smith, R. F., van den Bosch, R. and Hagen, K. S. (1959) The integration of chemical and biological control of the spotted alfalfa aphid (the integrated control concept). Hilgardia 29:81-101.

[17] Sharma, B.K. (1997) Environmental Chemistry, Krishna Ltd., Delhi pp. soil-114

[18] United States Protection Agency "Pesticides and Food: What Integrated Pest Management Means". 2011. http://www.epa.gov/pesticides/food/ipm.htm.

[19] United States Environmental Protection Agency "Integrated Pest Management (IMP) Principles". 2012. http://www.epa.gov/pesticides/factsheets/ipm.htm.

[20] van den Bosch, R and Stern, V. M. (1962) The Integration of Chemical and Biological Control of Arthropod Pests Annu. Rev. Entomol. 7: 367–387.

[21] Vereijken, P., Edwards, C., El Titi, A., Fougeroux, A. and Way., M. (1986) Proceedings of Workshop on Integrated Farming. Bull. I. O. B. C./W. P. R. S./IX 2, 1–34.

[22] Wikipedia, The Free Encyclopaedia http://en.wikipedia.org/wiki/Integrated_pest_management

[23] Wijnands, F. G. and Kroonen-Backbier, B. M. A. (1993) In Modern Crop Protection: Developments and Perspectives, ed. Zadoks, J. C. (Wageningen Pers, Wageningen, The Netherlands), pp. 227–234.

[24] Wright, L. 1996. Silent sperm. (A reporter at large.) The New Yorker (Jan. 15): 42,48,50-53,55.

[25] www.cias.wisc.edu/curriculum/.../Notes_for_IPM_powerpoint.doc

Chapter 7

Using Irrigation to Manage Weeds: A Focus on Drip Irrigation

Timothy Coolong

Additional information is available at the end of the chapter

1. Introduction

Irrigation has transformed agriculture and shaped civilization since its use in the Fertile Crescent more than 6000 years ago. Access to fresh water for irrigation transformed barren landscapes, allowing populations to move to previously uninhabitable regions. Advances in water management increased the productivity of agricultural systems around the world; supporting substantial population growth. Water consumption for agricultural use accounted for nearly 90% of global water use during the previous century [1] and is responsible for approximately 70% of fresh water withdrawals worldwide [2]. Currently, US water withdrawals for irrigation represent nearly 34% (137 billion gallons/day) of domestic water use [3]. Treating and pumping irrigation water has a significant carbon footprint as well. Pumping groundwater for irrigation requires about 150 kg Carbon/ha [4]. In the US more than 65% of total vegetable acreage and 76% of fruit acreage is irrigated [5]. Irrigating fruit and vegetable crops can increase marketable yields by 200% or more and is necessary to produce the high quality and yields required to be profitable [6]. It was estimated by Howell [5], that irrigated lands account for 18% of total cropped area, but produce approximately 50% of crop value. Due to the large observable increases in yield and quality associated with irrigation, many farmers tend to over-irrigate, viewing it as an insurance policy for growing fruits and vegetables. Irrigation can routinely exceed 10% of input costs in the US [7] and over-irrigating may reduce yields in some instances [8]. Excessive irrigation not only depletes freshwater reserves, but may leach fertilizers and other chemicals from agricultural lands [9-11]. Unnecessary applications of water and fertilizer can also allow weeds to flourish in modern agriculture. While irrigation systems are usually designed and managed with a crop of interest in mind; the impact of irrigation on weed growth is an important component of any mod-

ern production system. The following chapter will address the impact of different irrigation systems on weed management with an emphasis on drip irrigation technologies.

2. The influence of irrigation method on weeds

Surface, sprinkler, and drip irrigation are the three primary types of irrigation methods used
to grow crops (Figure 1). Within each method, there are several subcategories, each of which
varies in water use efficiency, cost, yield, and weed management potential.

Figure 1. An irrigation canal for furrow irrigation of cabbage (*Brassica oleracea*) (*left*), solid set sprinkler irrigation of
onion (*Allium cepa*) (*center*) and surface drip irrigation of recently seeded cabbage (*right*)

2.1. Surface irrigation and the impact on weeds

Surface irrigation, which floods entire fields or supplies water in furrows between planted
rows, is the most common type of irrigation used worldwide. Some surface irrigation systems have been operating continuously for thousands of years and have the ability to supply enormous quantities of water over widespread areas. Flood and furrow irrigation can
have water use efficiencies per unit of yield ranging from 25-50% of well managed drip irrigation systems [12]. One of the most common crops grown worldwide with flood irrigation
is lowland rice (*Oryza sativa*). Flood irrigation can be an integral part of weed management
for this crop.

As a semi-aquatic crop, lowland rice production utilizes substantial quantities of water. It
was estimated that more than 2m of water are used per crop of rice grown [13]. This under-
scores the substantial water requirements for lowland rice production; particularly in the initial flooding stages when large quantities of water may be lost prior to saturation [14].
Although it has been reported that rice grown under saturated field conditions did not experience additional water stress and yielded no differently than rice grown under standing
water [15,16]; rice which is grown under standing water competes better with weeds than
when grown in saturated soils [17, 18]. Although some weeds propagate vegetatively, most
develop from seeds; thus flooding can restrict the germination and reduce the abundance of
many weeds found in rice paddies [19].

Despite reducing the presence of some weed species, flooded lowland rice fields have over time selected for the presence of semi-or aquatic weed species. To reduce the presence of some of these weeds flooded soils are often tilled. While the primary goal of tillage is to up-root recently germinated weed seedlings; tilling flooded soils can destroy structure and po-rosity. This results in soils within low infiltration rates, which increases water retention, allowing fields to remain flooded [20].

Weed control in modern rice production is a system where irrigation management is inte-grated with tillage and planting practices as well as herbicides. Williams et al. [21] reported that weed control was better in fields submerged under 20 cm of water compared to those submerged under 5 cm of water when no herbicides were used. However, when herbicides were included weed control improved significantly at all depths [17]. Flowable-granular herbicide formulations, which are often used in lowland rice production, also rely on stand-ing water for dispersal. Flooded paddy fields allow uniform dispersal of low quantities of herbicides resulting in superior control of weeds [22, 23]. The integration of herbicides into the lowland rice production systems has reduced labor requirements for weed control by more than 80% since the introduction of 2,4-D in 1950, while simultaneously improving overall weed management [23]. Flooding has been an effective weed management technique in lowland rice for thousands of years. Coupled with modern herbicides, farmers can effi-ciently manage weeds on a large scale. Nonetheless, the high costs of water and demands on finite fresh water resources may result in substantial changes to the current lowland rice production system. The development of "aerobic-rice," drought tolerant lowland varieties that can yield well on non-saturated soils, may change how irrigation is used to manage weeds in lowland rice. Aerobic-rice is grown in a manner similar to many other grains, with land allowed to dry between irrigation cycles. This has the potential to reduce the reliance on flooding and irrigation water for weed control, likely shifting to chemical or mechanical methods [24].

Furrow irrigation is a common irrigation method where water is sent through ditches dug between raised beds to provide water to plants. Instead of flooding entire fields, only fur-rows between beds are wetted, allowing water to seep into growing beds through capillary action. Furrow irrigation is commonly used on millions of hectares of crops worldwide; where complex canal networks can move irrigation water hundreds of miles from upland sources to lower elevation growing areas. As would be expected, weed pressure in the irri-gated furrows between rows is generally higher than with the rows themselves [25]. To con-trol these weeds, mechanical cultivation may be used, but in many instances, herbicides, either applied to the soil as sprays or through irrigation water, are relied upon. However, the administration of herbicides through furrow irrigation can be challenging. Poor applica-tion uniformity, downstream pollution, and inaccuracies due to difficulties in measuring large quantities of water are challenges associated with applying herbicides through surface irrigation water [26, 27]. Chemical choice also is important when applying herbicides in sur-face irrigation systems. For example, Cliath et al. [27] noted that large quantities of the herbi-cide EPTC volatilized shortly after application via flood irrigation in alfalfa (*Medicago sativa*);

Amador-Ramirez et al. [28] reported variability in the effectiveness of some herbicides when applied through furrow irrigation compared to conventional methods.

A variant on the typical furrow irrigation system has been developed that combines furrow irrigation with polyethylene mulches and rainwater collection to irrigate crops, while controlling weeds. The production method, called the "ridge-furrow-ridge rainwater harvesting system," uses woven, water-permeable, polyethylene mulches that cover two ridges as well as a shallow furrow between the ridges [29, 30]. The system is similar to a raised-bed plastic mulch system, with inter-row areas being left in bare soil. However, unlike a traditional plastic mulch system, a furrow is made in the center of the raised bed to collect any rainwater that ordinarily would be lost as runoff from the bed. This system significantly reduces weed pressure in the furrow area and increases yield with the use of a polyethylene mulch, while reducing the need for supplemental irrigation by collecting rainwater [29]. Interestingly, a similar method of irrigation was employed during early experiments with plastic mulch, prior to the introduction of drip irrigation tubing. In these trials irrigation was achieved by cutting furrows in the soil next to the crop, covering them with plastic, and cutting holes in the plastic for the water to penetrate the plant bed [31, 32]

2.2. Sprinkler irrigation and the impact on weeds

Introduced on a large scale in the 1940s, sprinkler irrigation systems are used on millions of ha of crop land. The three primary types of sprinkler irrigation are center pivot, solid set, and reel or travelling gun systems. Sprinkler systems require a pump to deliver water at high pressures and are costlier than surface irrigation systems, but provide superior application uniformity and require less water to operate [33,34]. While center pivot systems require relatively level ground; solid set and reel-type systems can be used on with varied topographies. Because of improved application uniformity, sprinkler irrigation is the method of choice when applying herbicides or other agrichemicals through the irrigation system [26]. Sayed and Bedaiwy [35] noted a nearly 8-fold reduction in weed pressure when applying herbicides through sprinkler irrigation compared to traditional methods. Sprinkler irrigation permits growers to uniformly apply water over large areas, which can allow for proper incorporation of some preemergent herbicides [36]. In addition to applying herbicides, preplant sprinkler irrigation of fields, when combined with shallow tillage events after drying, has been shown to significantly reduce weed pressure during the growing season. This process of supplying water to weed seeds prior to planting, which causes them to germinate, where they can then be managed through shallow cultivation or through herbicide application is termed "stale seed-bedding" and is routinely used by farmers in many parts of the US.

2.3. Drip irrigation and the impact on weeds

Introduced on a large scale in the late 1960s and early 1970s, drip irrigation has steadily grown in popularity [37]. Although drip irrigation is only utilized on approximately 7% of the total irrigated acreage in the US, it is widely used on high value crops such as berries and vegetables [3]. Drip irrigation, if properly managed, is highly efficient with up to 95%

application efficiencies [38]. The productivity of drip irrigation has prompted significant increases (> 500%) in its use over the previous 20-30 years [5]. While drip irrigation is typically expensive and require significant labor to install and manage; the water savings compared to other methods of irrigation have prompted grower adoption. Drip irrigation has several benefits in addition to improved water use efficiencies. By only wetting the soil around plants leaves are kept dry reducing foliar disease and potential for leaf burn when using saline water [37, 39]. Fertilizers, which are easily supplied through drip irrigation, are restricted to an area near active rooting. This leads to more efficient use by the target crop. Because drip irrigation only wets the soil in the vicinity of the drip line or emitter, growers are able to supply irrigation water only in the areas required to grow the crop of interest. Soils between rows are not supplied with water or fertilizer, reducing weed growth. When drip irrigation is coupled with plastic mulch and preplant soil fumigation, weeds can be effectively controlled within rows, leaving only between-row areas to be managed. By restricting weed management to areas between rows growers increase their chemical and mechanical control options. While many farmers may apply preemergent herbicides to between-row areas, weeds that do germinate can be controlled easily with directed sprays of postemergent herbicides with low risk to the crops growing in the plastic mulch. In arid growing regions the combination of plastic mulch and drip irrigation may lead to acceptable weed control without the use of herbicides.

Because drip irrigation can supply limited quantities of water to an area immediately surrounding the crop root zone, it can be ideally suited for insecticide or fungicide injection. The small quantities of water delivered with drip irrigation requires significantly less chemical to maintain a given concentration applied to plants compared to surface or sprinkler irrigation [40]. However, while drip irrigation is one of the most efficient means to deliver chemicals such as systemic insecticides to plants, it is much less effective than comparable sprinkler systems for herbicide applications. The limited wetting pattern and low volume of water used for drip irrigation means that herbicides do not reach much of the cropped area. Within wetted areas herbicides may be degraded prior to the end of the season [26]. Because drip systems are often designed for frequent, low-volume irrigations, soils around plants may remain moist, reducing the efficacy of preemergent herbicides. Fischer et al. [41] reported significantly better weed control when using micro sprinklers compared to drip irrigation in vineyards and orchards. This was due to a reduction in the effectiveness of preemergent herbicides in drip irrigated treatments late in the growing season. The authors speculated that the drip irrigated plants had persistent soil moisture near the emitters resulting in enhanced degradation of the applied herbicides. Drip irrigation is often used in tandem with herbicides; however, they are often applied using conventional sprayers. Therefore, the weed control benefits of drip irrigation are due to the ability to precisely manage and locate water where it will most benefit crops while reducing availability for weed growth. One method that allows growers to precisely locate water in the root zone, below the soil surface, away from weed seeds is subsurface drip irrigation.

3. Subsurface drip irrigation

Subsurface drip irrigation (SDI) has been utilized in various forms for more than a century [37, 42]. Presently SDI uses standard drip irrigation tubing that is slightly modified for below-ground use. While typical surface drip irrigation tubing have walls that are usually 8 or 10-mil thick; tubing made specifically for multi-season SDI applications, have walls with a 15-mil thickness. In addition, tubing made specifically for SDI applications may have emitters which are impregnated with herbicides to prevent root intrusion [43]. Because growers are unable to inspect buried tubing, any problems with emitter clogging or cuts in the line may go unnoticed for long periods of time. Subsurface drip irrigation used for the production of high-value crops such as vegetables, which tend to have shallow root systems, may be buried at depths of 15-25 cm [44]. Subsurface drip tubing that is used for agronomic crops such as cotton (*Gossypium spp.*) or corn (*Zea maize*) is generally buried 40-50 cm below the soil surface [45]. Drip irrigation tubing used for agronomic crops is typically left in place for several years in order to be profitable and must reside below the tillage zone to avoid being damaged [45]. Agronomic crops in general tend to be deeper rooted than many vegetable crops allowing them to access water supplied at greater depths. In addition, the deeper placement of the irrigation tubing reduces the potential rodent damage, which can be significant [45, 46].

Drip tubing may be placed during or after bed formation in tilled fields or into conservation tillage fields with drip tape injection sleds (Figure 2). While SDI that is used for a single season may be connected to flexible "lay-flat" tubing at the ends of fields; more permanent installations are generally coupled to rigid PVC header lines.

Figure 2. Injection sled for SDI.

Although concern over buried drip tubing collapsing under the pressure of the soil above is justified; properly maintained SDI systems have lasted 10-20 years in the Great Plains without significant problems [45]. For permanent systems, lines must be cleaned and flushed after every crop if not more frequently. In single-season trials conducted by the author, end of season flow rates were found to be no different between surface and SDI systems placed at a depth of 15 cm (*T. Coolong, unpublished data*). However, when comparing SDI that had been in use for three years for onion production to new SDI tubing, there were slight reductions in discharge uniformity in the used tapes [47].

3.1. Subsurface drip irrigation in organic farming

Some of the earliest uses of SDI were not based on enhanced water use efficiency but because drip irrigation tubing on the soil surface could interfere with agricultural equipment, particularly cultivation tools [48]. While many conventional farmers now rely more on chemical weed control than on cultivation, most organic growers must rely exclusively on cultivation to manage weeds. For this reason, SDI is particularly appropriate for organic farming systems. Traditional placement of drip irrigation tubing requires growers to remove the tubing prior to cultivation, increasing labor costs. By burying drip tubing below the depth of cultivation, growers can control weeds mechanically. SDI is routinely used for bare-ground, organic vegetable production at The University of Kentucky Center for Horticulture Research (Lexington, KY, US). This system uses a SDI injection sled (Figure 2) coupled with in-row cultivators to effectively control weeds in a humid environment (Figure 3).

Figure 3. Buried drip irrigation tubing entering the soil at the end of a field (top left), a two-row cultivator using side-knives and spring hoes (top right), a rolling basket weeder controlling weeds within and between rows (bottom left), and organically-managed kale and collard *(Brassica oleracea* Acephala group) crops (bottom right) that are grown with SDI and mechanical cultivation for near complete weed control in a humid environment.

In this system, SDI tubing is placed approximately 15 cm below the surface on a shallow raised bed. Using SDI in combination with precision cultivation has allowed for nearly complete control of weeds on an organic farm in an environment which may regularly experience 25 cm or more rain during the growing season.

3.2. Subsurface drip irrigation and water use

More than 40 types of crops have been tested under SDI regimes [42]. In most cases yields with SDI were no different than or exceeded yields for surface drip irrigation. In many cases water savings were substantial. However, SDI relies on capillary movement of water upward to plant roots. Soil hydraulic properties can significantly affect the distribution patterns of water around emitters, making interpretation of data difficult when comparing the effectiveness of SDI in different soil types [49]. Trials often report water savings or increased yield in SDI systems compared to surface drip systems for vegetable crop production [44, 50, 51], although some do not [46].

In 2012, studies were conducted at The University of Kentucky Center for Horticulture Research (Lexington, KY, US) comparing SDI at a depth of 15 cm to surface placement of drip irrigation tubing for the production of acorn squash (*Cucurbita pepo*) 'Table Queen'. The soil was a Maury silt loam series, mesic Typic Paleudalfs. Irrigation was controlled automatically with switching-tensiometers placed at a depth of 15 cm from soil surface [52, 53]. Tensiometers were placed approximately 20 cm from plants and 15 cm from the drip tubing which was centered on raised beds. Tensiometer set points were as follows: on/off -40/-10 kPa and -60/-10 kPa for both SDI and surface drip systems. In both moisture regimes the surface applied drip irrigation utilized less water during the growing season than SDI (Table 1). Interestingly, the number of irrigation events and the average duration of each event varied significantly among the surface and SDI treatments when irrigation was initiated at -40 kPa, but were similar when irrigation was scheduled at -60 kPa. Irrigations were frequent, but relatively short for the -40/-10 kPa surface irrigation treatment. Comparable results have been reported in studies conducted in tomato (*Lycopersicon esculentum* syn. *Solanum lycopersicum*) and pepper (*Capsicum annuum*) using a similar management system and set points. However the SDI -40/-10 kPa treatment irrigated relatively infrequently and for longer periods of times. When irrigation was initiated at -60 kPa and terminated at -10 kPa there were differences in water use between the two drip systems, with the surface system being more efficient. However, unlike the -40/-10 kPa treatments, the numbers of irrigation events were not different between the two drip irrigation systems. The difference in the response of the SDI and surface systems when compared under different soil moisture regimes was not expected and suggests that irrigation scheduling as well as soil type may have a significant impact on the relative performance of SDI compared to surface drip irrigation. This should be noted when comparing the performance of SDI and surface drip irrigation systems.

3.3. Subsurface drip irrigation for improved weed management

As previously discussed, a key benefit of SDI is a reduction in soil surface wetting for weed germination and growth. Although the lack of surface wetting can negatively impact direct-seeded crops, transplanted crops often have significant root systems that may be wetted without bringing water to the soil surface. Direct-seeded crops grown with

SDI are often germinated using overhead microsprinkler irrigation [51]. The placement of SDI tubing as well as irrigation regime [54] can impact the potential for surface wetting and weed growth. As mentioned previously, SDI is often located 40-50 cm below the soil surface in most agronomic crops, but is typically shallower (15-25 cm) for vegetable crops [51]. Patel and Rajput [55] evaluated five depths (0, 5, 10, 15, and 20 cm) of drip irrigation with three moisture regimes in potato (*Solanum tuberosum*). Soil water content at the surface of the soil was relatively moist for drip tubing placed 5 cm below the surface, while the soil surface remained relatively dry for the 10, 15, and 20 cm depths of drip tubing placement [55]. Because that study was carried out on sandy (69% sand) soils, greater depths may be required to prevent surface wetting on soils with a higher clay content and greater capillary movement of water [56].

Irrigation treatment on/off	Irrigation type	Events	Mean irrigation time	Mean irrigation vol.
kPa		no.	min/event	l•ha⁻¹
-40/-10	Surface	48	92	1.25 x 10⁶
-40/-10	SDI	18	276	1.50 x 10⁶
-60/-10	Surface	14	201	0.84 x 10⁶
-60/-10	SDI	14	251	1.06 x 10⁶

Mean number of irrigation events, irrigation time per event, and irrigation volume for the season 'Table Queen' squash grown with automated irrigation in 2012 in Lexington, KY.

Table 1. A comparison of SDI and surface drip irrigation under two automated irrigation schedules.

SDI not only keeps the soil surface drier, but also encourages deeper root growth than surface drip systems. Phene et al. [57] reported greater root densities below 30 cm in sweet corn grown under SDI compared to traditional surface drip. In that study the SDI tubing was placed at a depth of 45 cm. In bell pepper, a shallow rooted crop, SDI encouraged a greater proportion of roots at depths below 10 cm when laterals were buried at 20 cm [58]. Encouraging deeper root growth may afford greater drought tolerance in the event of irrigation restrictions during the production season.

In arid climates SDI has been shown to consistently reduce weed pressure in several crops, including cotton, corn, tomato, and pistachio (*Pistacia vera*) [25, 59, 60]. For example, weed growth in pistachio orchards in Iran was approximately four-fold higher in surface irrigated plots compared to those with SDI [59]. In humid regions, benefits may depend on the level of rainfall received during the growing season; however, a reduction in the consistent wetting of the soil surface should allow for a reduction in weed pressure, particularly when coupled with preemergent herbicides (Figure 4).

Processing tomatoes represent one of the most common applications of SDI in vegetable crops. The impact of SDI (25 cm below the soil surface) and furrow irrigation on weed growth were compared in tomato [25]. In that trial the authors reported a significant decrease in weed growth in plant beds and furrows with SDI compared to furrow irrigation. When no herbicides were applied, annual weed biomass was approximately 1.75 and 0.05 tons per acre dry weight in the furrow and SDI treatments, respectively [25]. With herbicides, both irrigation treatments had similar levels of weed biomass. However, in that study, weed biomass in the SDI non-herbicide treatment was similar to the furrow irrigation with herbicide treatment, suggesting that when using SDI, herbicides may not be necessary in arid environments.

Figure 4. The difference in weed growth approximately 10 days after transplanting between acorn squash (*Cucurbita pepo*) which were subjected to SDI at a depth of 15 cm below the soil surface (*left*) and surface drip irrigation (*right*). A preemergent herbicide (halosulfuron methyl, Sandea™) was applied to all plots

A similar trial compared SDI and furrow irrigation across different tillage regimes with and without the presence of herbicides in processing tomato [60]. In that study, both conservation tillage and SDI reduced the weed pressure compared to conventional alternatives. However, when main effects were tested, SDI had the largest impact on weed growth of any treatment. Main effects mean comparisons showed that SDI treatments had weed densities of 0.5 and 0.6 weeds per m^2 in the planting bed in years one and two of the trial, respectively, compared to 17.9 and 98.6 weeds per m^2 in the plant bed for furrow irrigated treatments. As would be expected, SDI substantially reduced weed populations in the furrows between beds as they remained dry during the trial. In this trial SDI had a greater impact on weed populations than herbicide applications. The authors concluded that SDI could reduce weed populations sufficiently in conservation tillage tomato plantings in arid environments such that herbicides may not be necessary [60]. In another related trial, weed populations were evaluated for processing tomatoes grown with SDI and furrow irrigation under various weed-management and cultivation systems [46]. In this trial, the authors noted an increase in weed densities in the furrow system compared to SDI within the planting bed and furrows. However, there was no significant difference in the total weed biomass in the plant bed comparing the two irrigation systems [46]. The authors did note that the majority of the weeds in the SDI treatment were in the plant row and not evenly distributed across the bed, indicating that the outer regions of the plant bed were too dry to support weed germination or growth. Interestingly, when the relative percentages of weeds are extrapolated from the results provided, *Solanum nigrum* constituted 76% of the weed population in the plant beds of SDI treatments, but 52% of the weed population in the furrow irrigated beds. Although the sample size from that study is too small to make statements regarding selection pressures for weed species in the irrigation systems evaluated, it may give insight into why the authors reported a significant difference in numbers of weeds, but not biomass. *Solanum nigrum* can grow quite large and may have contributed a substantial amount of biomass in the SDI plots, despite having fewer numbers of weeds present. In this trial the furrow irrigation treatments had significantly greater yields than the SDI treatments [46]. The authors suggested that this was not due to a flaw in the SDI system, but poor management late in the season. The relatively small amounts of water used in drip irrigation underscore the need for proper scheduling; otherwise water deficits can occur, resulting in poor yields.

4. Efficient management of drip irrigation

Appropriate management of irrigation requires growers to determine when and how long to irrigate. A properly designed and maintained drip irrigation system has much higher application efficiencies than comparable sprinkler or surface irrigation systems [37]. However, even with drip irrigation, vegetable crops can require large volumes of water - more than 200,000 gallons per acre for mixed vegetable operations in Central Kentucky, US [61]. Poorly managed drip irrigation systems have been shown to reduce yields (Locascio et al., 1989) and waste significant quantities of water. Just 5 h after the initiation of drip irrigation, the wetting front under an emitter may reach 45 cm from the soil surface, effectively below the root zone of many vegetables [62]. When drip irrigation is mismanaged, a key benefit – limiting water available for weeds, is lost. The ability to precisely apply water with drip irrigation means that a very high level of management can be achieved with proper scheduling [63].

Irrigation scheduling has traditionally been weather or soil-based; although several plant-based scheduling methods have been proposed [64, 65]. In weather-based scheduling, the decision to irrigate relies on the soil-water balance. The water balance technique involves determining changes in soil moisture over time based on estimating evapotranspiration (Et) adjusted with a crop coefficient [66]. These methods take environmental variables such as air temperature, solar radiation, relative humidity and wind into account along with crop coefficients that are adjusted for growth stage and canopy coverage [64]. Irrigating based on Et can be very effective in large acreage, uniformly planted crops such as alfalfa, particularly when local weather data is available. However, irrigating based on crop Et values for the production of vegetable crops is prone to inaccuracies due to variations in microclimates and growing practices. Plastic mulches and variable plant spacing can significantly alter the accuracy of Et estimates [67, 68]. Furthermore the wide variability observed in the growth patterns in different cultivars of the same vegetable crop can substantially alter the value of crop coefficients at a particular growth stage. In many regions of the US, producers do not have access to sufficiently local weather data and the programs necessary to schedule irrigation.

An alternative to using the check-book or Et-based models for irrigation is to use soil moisture-based methods. Perhaps the simplest and most common method is the "feel method," where irrigation is initiated when the soil "feels" dry [69]. Experienced growers may become quite efficient when using this method. More sophisticated methods of scheduling irrigation may use a tensiometer or granular matrix type sensor [6, 70-72].

These methods require routine monitoring of sensor(s), with irrigation decisions made when soil moisture thresholds have been reached. This requires the development of threshold values for various crops and soil types. Soil water potential thresholds for vegetable crops such as tomato and pepper have been developed [6, 72, 73]. Drip irrigation is well suited to this type of management as it is able to frequently irrigate low volumes of water allowing growers to maintain soil moisture at a near constant level [6, 52, 53, 72]. In some soils, high-frequency, short-duration irrigation events can reduce water use while maintaining yields of tomato when compared to a traditionally scheduled high-volume, infrequent irrigation (Table 2) [52, 71].

| Irrigation on/off | 2009 | | | 2010 | | |
	Events	Mean irrigation time	Mean irrigation vol.	Events	Mean irrigation time	Mean irrigation vol.
kPa	no.	min/event	l•ha^{-1}	no.	min/event	l•ha^{-1}
-30/-10	39	110	1.30 x 10^6	28	144	1.22 x 10^6
-30/-25	59	91	1.63 x 10^6	22	140	0.93 x 10^6
-45/-10	21	221	1.41 x 10^6	22	167	1.11 x 10^6
-45/-40	76	40	0.92 x 10^6	18	146	0.79 x 10^6

Mean number of irrigation events, irrigation time per event, and irrigation volume for the season 'Mountain Fresh' tomato grown under five automated irrigation regimes in 2009 and 2010 in Lexington, KY.

Table 2. A comparison of high frequency short duration to more traditional infrequent but long duration irrigation scheduling using soil moisture tension to schedule irrigation (Adapted from [52])

Coolong et al. [52] reported that irrigation delivered frequently for short durations so as to maintain soil moisture levels in a relatively narrow range could save water and maintain yields, but efficiencies varied depending on season and the soil moisture levels that were maintained. In two years of trials, irrigation water was most efficiently applied when soil moisture was maintained between -45 and -40 kPa for tomatoes grown on a Maury Silt Loam soil. However, when soil moisture was maintained slightly wetter at -30 to -25 kPa, the relative application efficiency was affected by growing year (Table 2). Therefore, while an effective method, soil moisture-based irrigation scheduling may produce variable application efficiencies and should be used in concert with other methods.

After more than 40 years of research with drip irrigation, results suggest that a mix of scheduling tactics should be employed to most efficiently manage irrigation. The application efficiencies of several different management methods were determined by DePascale et al. [12]. Those authors estimated that when compared to a simple timed application, the use of soil moisture sensors to schedule irrigation would increase the relative efficiency of drip irrigation by 40-50%. Using a method incorporating climate factors and the water-balance technique, one could increase relative efficiency compared to the baseline by 60-70%. However, when soil moisture sensors were combined with Et-based methods, the relative efficiency of drip irrigation could be increased by more than 115% over a fixed interval method. Therefore multiple strategies should be used to optimize drip irrigation scheduling. This ensures maintaining yields while reducing excessive applications of water, reducing the potential for weed growth.

5. Conclusions

Irrigation management is essential to developing a holistic system for weed management in crops. As water resources become costlier, drip irrigation technologies will become more widely utilized by growers worldwide. Although drip irrigation may be adopted due to water savings, the impact of drip irrigation on weed control is noteworthy. The ability to reduce soil wetting will allow for improved weed control over sprinkler and surface irrigation systems. Furthermore, precisely locating water in the root-zone without wetting the soil surface will make SDI more attractive to growers, despite the higher installation costs. In addition, SDI is now being implemented on large acreages for the production of grain crops, particularly corn, in the Midwestern US. With the increase in adoption of SDI, new technologies will be developed to overcome some of the limitations of that system. Future research will likely continue to develop management tactics combining multiple scheduling strategies such as Et and soil moisture-based irrigation [12] and its application for managing SDI on a wider range of crops and soil types.

Author details

Timothy Coolong

Department of Horticulture, University of Kentucky, USA

References

[1] Shiklomanov, IA. Appraisal and assessment of world water resources. Water International 2000; 25(1):11-32.

[2] Siebert S., Burke J., Faures JM., Frenken K., Hoogeveen J., Doll P., Portmann FT. Groundwater use for irrigation – a global inventory. Hydrology and Earth System Sciences Discussions 2010; 14:1863-1880.

[3] Hutson SS., Barber NL., Kenny JF., Linsey KS., Lumia DS., Maupin MA. Estimated use of water in the United States in 2000. Reston, Virginia: US Geological Survey; 2004.

[4] Lal R. Carbon emissions from farm operations. Environment International 2004; 30(7):981-990.

[5] Howell TA. Enhancing water use efficiency in irrigated agriculture. Agronomy Journal; 2001; 93(2):281-289.

[6] Smajstrla AG., Locascio SJ. Tensiometer – controlled, drip irrigation scheduling of tomato. Applied Engineering 1996; 12(3):315-319.

[7] Hochmuth GJ., Locascio SJ., Crocker TE., Stanley CD., Clark GA., Parsons L. Impact of microirrigation on Florida horticulture. HortTechnology 1993; 3(2):223-229.

[8] Locascio SJ., Olson SM., Rhoads FM. Water quantity and time of N and K application for trickle irrigated tomatoes. Journal of the American Society for Horticulture Science 1989; 114(2):265-268.

[9] Correll DL. The role of phosphorus in the eutrophication of receiving waters: a review. Journal of Environmental Quality 1998; 27(2):261-266.

[10] Hallberg GR. Pesticide pollution of groundwater in the humid United-States. Agriculture Ecosystems and Environment 1989; 26(3-4):299-367.

[11] Tilman D., Fargione J., Wolff B., D'Antionio C., Dobson A., Howarth R., Schindler D., Schlesinger WH., Simberloff D., Swackhamer D. Forecasting agriculturally driven global environmental change. Science 2001; 292(5515):281-284.

[12] De Pascale S., Dalla Costa L., Vallone S., Barbieri G., Maggio A. Increasing water use efficiency in vegetable crop production: from plant to irrigation. HortTechnology 2011; 21(3):301-308.

[13] Bhuiyan SI., Sattar MA., Khan AK. Improving water use efficiency in rice irrigation through wet seeding. Irrigation Science 1995; 16(1):1-8.

[14] Valera A. Field studies on water use and duration for land preparation for lowland rice. MS Thesis. University of Philippines; 1976.

[15] Bhuiyan SI. Irrigation system management research and selected methodological issues. International Rice Research Institute Research Paper Series 81. Los Banos, Phillipines International Rice Research Institute; 1982.

[16] Tabbal DF., Lampayan RM., Bhuiyan SI. Water efficient irrigation technique for rice. In: Proceedings of the International Workshop on Soil and Water Engineering for Paddy Field Management, 28-30 January 1992, Bangkok, Thailand. Asian Institute of Technology; Pathumthani, Thailand; 1992.

[17] Bhagat RM., Bhuiyan SI., Moody K. Water, tillage and weed interactions in lowland rice: a review. Agricultural Water Management 1996; 31(3):165-184.

[18] Matsunaka S. Evolution of rice weed control practices and research: world perspective. In: Swaminathan A. (ed.) Weed Control in Rice. Los Banos, Phillippines: International Rice Research Institute; 1983. p5-17.

[19] Zimidahl RL., Moody K., Lubigan RT., Castin EM. Patterns of weed emergence in tropical soil. Weed Science 1988; 36(5):603-608.

[20] Sharma PK., De Datta SK. Physical properties and processes of puddled rice soils. In: Stewart BA. (ed.) Advances in Soil Science 5. Berlin: Springer; 1986. p139-178.

[21] Williams JF., Roberts SR., Hill JE., Scardaci SC., Tibbits G. Managing water for weed control in rice. California Agriculture 1990; 44(5):7-10.

[22] Kamoi M., Noritake K. Technical innovation in herbicide use. Japan International Research Center for Agricultural Sciences IRCAS International Symposium Series 1996; 4:97-106.

[23] Watanabe H. Development of lowland weed management and weed succession in Japan. Weed Biology and Management 2011; 11(4):175-189.

[24] Tuong TP., Bouman BAM. Rice production in water scarce environments. In: Kijne JW., Barker R., Molden D. (eds.) Water productivity in agriculture: limits and opportunities for improvement. Wallingford, UK: CABI; 2003. p53-67.

[25] Grattan SR., Schwankl LJ., Lanini WT. Weed control by subsurface drip irrigation. California Agriculture 1988; 42(3): 22-24.

[26] Ogg AG. Applying herbicides in irrigation water - a review. Crop Protection 1986; 5(1):53-65.

[27] Cliath MM., Spencer WF., Farmer WJ., Shoup TD., Grover R. Volatilization of s-ethyl N,N-dipropylthiocarbamate from water and wet soil during and after flood irrigation of an alfalfa field. Journal of Agricultural and Food Chemistry 1980; 28(3): 610-613.

[28] Amador-Ramirez MD., Mojarro-Davila F., Velasquez-Valle R. Efficacy and economics of weed control for dry chile pepper. Crop Protection 2007; 26(4):677-682.

[29] Gosar B., Baricevic D. Ridge furrow ridge rainwater harvesting system with mulches and supplemental irrigation. HortScience 2011; 46(1):108-112.

[30] Li XY., Gong LD., Gao QZ., Li FR. Incorporation of ridge and furrow method of rainfall harvesting with mulching for crop production under semiarid conditions. Agricultural Water Management 2001; 50(3):173-183.

[31] Emmert, EM. Polyethylene mulch looks good for early vegetables. Market Growers' Journal 1956; 85:18-19.

[32] Emmert, EM. Black polyethylene for mulching vegetables. Proceedings of the American Society for Horticultural Science 1958; 69:464-469.

[33] Locascio SJ. Management of irrigation for vegetables: past, present, and future. HortTechnology 2005; 15(3):482-485.

[34] Sammis TW. Comparison of sprinkler, trickle, subsurface, and furrow irrigation methods for row crops. Agronomy Journal 1980; 72(5):701-704.

[35] Sayed MA., Bedaiwy MNA. Effect of controlled sprinkler chemigation on wheat crop in a sandy soil. Soil and Water Research 2011; 6(2):61-72.

[36] Dowler CC. Advantages of herbigation. International Water and Irrigation Review 1995; 15(3):26-29.

[37] Dasberg S., Or D. Drip Irrigation. Berlin: Springer; 1999.

[38] Rogers DH., Lamm FR., Alam M., Trooien TP., Clark GA., Barnes PL., Mankin K. Efficiencies and water losses of irrigation systems. Irrigation Management Series MF-2243. Manhattan, Kansas: Kansas State University Cooperative Extension Service; 1997.

[39] Yarwood CE. Water and the infection process. In: Kozlowsky TT. (ed.) Water Deficits and Plant Growth, Volume 5. New York: Academic Press; 1978. p141-156.

[40] Clark GA., Smajstrla AG. Injecting chemicals into drip irrigation systems. HortTechnology 1996; 6(3):160-165.

[41] Fischer BB., Goldhamer DA., Babb T., Kjelgren R. Preemergence herbicide performance under drip and low volume sprinkler irrigation. In: Howell TA. (ed.) Drip/Trickle Irrigation in Action Volume II: Proceedings of the Third International Drip/Trickle Irrigation Congress, 18-21 November 1985, Fresno California. St. Joseph, Michigan: American Society of Agricultural Engineers; 1985.

[42] Camp CR. Subsurface drip irrigation: a review. Transactions of the ASAE 1998; 41(5): 1353-1367.

[43] Zoldoske DF., Genito S., Jorgensen GS. Subsurface drip irrigation (SDI) on turfgrass: a university experience. In: Lamm FR. (ed.) Microirrigation for a Changing World: Conserving Resources/Preserving the Environment: Proceedings of the Fifth Interna-

tional Microirrigation Congress, 2-6 April 1995, Orlando, Florida. St. Joseph, Michigan: American Society of Agricultural Engineers; 1995.

[44] Zotarelli L., Scholberg JM., Dukes MD., Munoz-Carpena R., Icerman J. Tomato yield, biomass accumulation, root distribution and irrigation water use efficiency on a sandy soil, as affected by nitrogen rate and irrigation scheduling. Agricultural Water Management 2009; 96(1):23-34.

[45] Lamm FR., Trooien TP. Subsurface drip irrigation for corn production: a review of 10 years of research in Kansas. Irrigation Science 2003; 22(3-4):195-200.

[46] Shrestha A., Mitchell JP., Lanini WT., Subsurface drip irrigation as a weed management tool for conventional and conservation tillage tomato (Lycopersicon esculentum Mill.) production in semi-arid agroecosystems. Journal of Sustainable Agriculture 2007; 31(2):91-112.

[47] Safi B., Neyshabouri MR., Nazemi AH. Water application uniformity of a subsurface drip irrigation system at various operating pressures and tape lengths. Turkish Journal of Agriculture and Forestry 2007; 31(5):275-285.

[48] Tollefson S. Subsurface drip irrigation of cotton and small grains. In: Howell TA. (ed.) Drip/Trickle Irrigation in Action Volume II: Proceedings of the Third International Drip/Trickle Irrigation Congress, 18-21 November 1985, Fresno California. St. Joseph, Michigan: American Society of Agricultural Engineers; 1985.

[49] Lazarovitch N., Shani U., Thompson TL., Warrick AW. Hydraulic properties affecting dishcharge uniformity of gravity-fed subsurface drip irrigation systems. Journal of Irrigation and Drainage Engineering 2006; 132(6):531-536.

[50] Beyaert RP., Roy RC., Ball-Coelho BR. Irrigation and fertilizer management effects on processing cucumber productivity and water use efficiency. Canadian Journal of Plant Science 2007; 87(2):355-363.

[51] Lamm FR., Camp CR. Subsurface Drip Irrigation. In: Lamm FR., Ayars JE., Nakayama FS. (eds.) Microirrigation for Crop Production. Amsterdam: Elsevier; 2007. p473-551.

[52] Coolong T., Surendran S., Warner R. Evaluation of irrigation threshold and duration for tomato grown in a silt loam soil. HortTechnology 2011; 21(4):466-473.

[53] Coolong T., Snyder J., Warner R., Strang J., Surendran S. The relationship between soil water potential, environmental factors and plant moisture status for poblano pepper grown using tensiometer-scheduled irrigation. International Journal of Vegetable Science 2012; 18(2):137-152.

[54] Hanson B., May D. Effect of subsurface drip irrigation on processing tomato yield, water table depth, soil salinity, and profitability. Agricultural Water Management 2004; 68:(1)1-17.

[55] Patel N., Rajput TBS. Effect of drip tape placement depth and irrigation level on yield of potato. Agricultural Water Management 2007; 88(1-3):209-223.

[56] Jury WA., Horton R. Soil Physics 6th Edition. Hoboken: Wiley; 2006.

[57] Phene CJ., Davis KR., Hutmacher RB., Bar-Yosef B., Meek DW., Misaki J. Effect of high frequency surface and subsurface drip irrigation on root distribution of sweet corn. Irrigation Science 1991; 12(3):135-140.

[58] Kong Q., Li G., Wang Y., Huo H. Bell pepper response to surface and subsurface drip irrigation under different fertigation levels. Irrigation Science 2012; 30(3):233-245.

[59] Dastorani MT., Heshmati M., Sadeghzadeh MA. Evaluation of the efficiency of surface and subsurface irrigation in dryland environments. Irrigation and Drainage 2010; 59(2):129-137.

[60] Sutton KF., Lanini WT., Mitchell JP., Miyao EM., Shrestha A. Weed control, yield, and quality of processing tomato production under different irrigation, tillage, and herbicide systems. Weed Technology 2006; 20(4):831-838.

[61] Spalding D. On-farm commercial vegetable demonstrations. In Coolong T., Snyder J., Smigell S. (eds.) 2008 Fruit and Vegetable Research Report. Lexington, Kentucky: University of Kentucky Cooperative Extension Service; 2008. p11-12.

[62] Elmaloglou S., Diamantopoulos E. Wetting front advance patterns and water losses by deep percolation under the root zone as influenced by pulsed drip irrigation. Agricultural Water Management 2007; 90(1-2):160-163.

[63] Hartz TK. Water management in drip-irrigated vegetable production. HortTechnology 1996; 6(3):165-167.

[64] Fereres E., Golhamer DA., Parsons LR. Irrigation water management of horticultural crops. HortScience 2004; 38(5):1036-1042.

[65] Jones HG. Irrigation scheduling: advantages and pitfalls of plant-based methods. Journal of Experimental Botany 2004; 55(407):2427-2436.

[66] Penman HL. Natural evaporation from open water, bare soil and grass. Proceedings of the Royal Society of London Series A- Mathematical and Physical Sciences 1948; 193(1032):120-145.

[67] Amayreh J., Al-Abed N. Developing crop coefficients for field-grown tomato (Lycopersicon esculentum Mill.) under drip irrigation with black plastic mulch. Agricultural Water Management 2005; 73(3):247-254.

[68] Burman RD., Nixon PR., Wright JL., Pruitt WO. Water requirements. In: Jensen ME. (ed.) Design and Operation of Farm Irrigation Systems. St. Joseph, Michigan: American Society of Agricultural Engineers; 1980. p189-232.

[69] Maynard DN., Hochmuth GJ. Knott's Handbook for Vegetable Growers Fifth Edition. Hoboken: Wiley; 2007.

[70] Richards LA., Russell MB., Neal OR. Further Developments on Apparatus for Field Moisture Studies. Proceedings of the Soil Science Society of America 1938; 2:55-64.

[71] Munoz-Carpena R., Dukes MD., Li YCC., Klassen W. Field comparison of tensiometer and granular matrix sensor automatic drip irrigation on tomato. HortTechnology 2005; 15(3):584-590.

[72] Thompson RB., Gallardo M., Valdez LC., Fernandez MD. Using plant water status to define threshold values for irrigation management of vegetable crops using soil moisture sensors. Agricultural Water Management 2007; 88(1-3):147-158.

[73] Smittle Da., Dickens WL., Stansell JR. Irrigation regimes affect yield and water use by bell pepper. Journal of the American Society for Horticultural Science 1994; 199(5): 936-939.

Chapter 8

Soil Physical Quality and Carbon Stocks Related to Weed Control and Cover Crops in a Brazilian Oxisol

Cezar Francisco Araujo-Junior,
Benedito Noedi Rodrigues,
Júlio César Dias Chaves and
George Mitsuo Yada Junior

Additional information is available at the end of the chapter

1. Introduction

The integration of weed management and cover crops as green manure plays an important role in several soil physical processes, influences carbon stocks and might be useful for mini-mizing soil physical degradation by compaction and hydric erosion.

According to Lal [1], soil degradative processes can be divided into physical (soil structure deg-radation, leading to crusting, compaction, erosion, desertification, anaerobiosis, environmen-tal pollution and unsustainable use of natural resources), chemical and biological. Soil structure is an important soil physical property that affects all three degradative processes [1] and might be changed by weed control and cover crops management.

In an early report, Rufino et al. [2] investigated the relationships between management of a coffee crop and bare soil during five years on a Dystropherric Red Latosol with a 6% slope at Londrina. They observed that the soil losses in the coffee plantation were similar to bare soil (99.30 Mg ha^{-1}) in the first and second year post coffee planting. However, they noted that the soil losses decreased to 33.93 Mg ha^{-1} from the fourth to fifth year. It was inferred that increasing soil cover between coffee rows and under the coffee canopy is important in reducing erosion susceptibility.

In another study conducted at Londrina, it was shown that high coffee population densities resulted in an increase in soil water content, soil organic matter, soil nutrient availability and a decrease in soil acidity and concentration of carbon in the soil microbial biomass (Pavan et al. [3]). They suggested that these results were due to the greater quantity of residues on the soil surface with higher coffee population density, which increased soil water content in both assessed layers.

In a Typical Dystropherric Red Latosol (617 g kg^{-1} clay content) with 12% slope, Carvalho et al. [4] quantified the lowest soil losses (0.1098 Mg ha^{-1} ano^{-1}) and sediment transportation (0.025 Mg ha^{-1} mm$^{-1)}$ in a system where the weed control was mechanical mowing. This obtained greater protection against soil erosion than, when the weed control was hand hoeing and the soil was exposed, increasing soil losses to 67.2434 Mg ha^{-1} ano^{-1} and sediment transportation to 0.022 Mg ha^{-1} mm^{-1}. They also noted that weed control with post-emergence herbicide had an intermediate effect in relation to soil loss and sediment transportation.

Faria et al. [5] observed that, in bare soil, combined application of pre-emergence herbicide and systemic herbicide showed clear signs of surface crusting and sheet erosion associated with the formation of micro-rills and micro-knolls on the surface. As a consequence of this surface crusting there were increases in soil strength, quantified by precompression stress and soil load bearing capacity [6, 7].

Some studies carried out in tropical Oxisols and Ultisols [4 — 9] have shown the effect of weed control on soil physical, mechanical and biological properties. As reported earlier, in coffee plantations in the State of Paraná intense cultivation of coffee resulted in severe declines in soil organic matter contents and the use of large heavy farm equipment has produced compacted soils with poor structure that are susceptible to erosive rainfall [10]. Thus, cover crops like dwarf mucuna and peanut horse planted between coffee rows might be useful to decrease soil susceptibility to hydric erosion.

Dwarf mucuna [*Mucuna deeringiana* (Bort.) Merr] is a tropical legume and among the most successful species for using as a cover crop or green manure between coffee rows. It is a shrubby species of determinate growth, has a short or early cycle, and reaches a maximum height of about 40 to 50 cm [11]. Furthermore, the production of plant biomass between 4 – 6 Mg ha^{-1}, which minimizes the severe damage caused by water erosion, improves the root system by the decomposition of crop residues, reduces the time spent in the management of weeds, increases production and improves the nutrition of the coffee plants [12], consequently decreasing the cost of coffee production by decreasing fertilizer dependence.

Peanut horse [*Arachis hypogea*] is a legume with a long cycle (200 days between sowing to harvest). This ensures good coverage and protection of the soil [13] during all the periods of most intense rainfall (between October to February) [14], when the rainfall causes high erosion [15]. Furthermore, planting this crop between the lines of coffee favours biological fixation, increasing the cycling of nutrients and dry matter production to between 2200 and 2550 kg per hectare, which contributes to the maintenance of the soil moisture [13].

Furthermore, the impact of weed management on the total organic carbon concentration in soil might be affected per unit area or volume increase, as well as soil bulk density and thickness of soil layer [16].

The soil-water retention curve defines the relationship between the matrix pressure head and water content [17, 18, 19, 20]. Any soil-water retention curve has certain common features that reflect the forces influencing the water retention [21]. Soil structure might influence these forces and change the behaviour of the soil-water retention curve.

The distinguishing properties of the soil-water retention curves depend on several factors, such as soil structure and aggregation, initial moulding water content, void ratio, type of soil, particle size distribution, mineralogy, stress history and soil compaction state. Among these factors, the stress history and initial moulding water content have the most influence on soil structure, which in turn dominates the nature of the soil-water retention curve and governs the air-entry value [18]. Authors have also shown that sandy clay till soil has two levels of structure: a macro-level structure and a micro-level structure and that both levels of structure are present in natural and compacted clayey soils.

Dexter [22] proposed the slope "S" of the soil-water retention curve at the inflection point as a measure of the micro-structural porosity of the soil for assessment of soil physical quality. This author also showed that the S-index was related to particle size distribution, soil bulk density, soil organic matter and root growth. This index, according to Dexter [22], is mostly due to microstructural porosity, and therefore, S governs many of these principal soil physical properties directly.

Although, some studies done in tropical soils suggest that the same reference value of S-index might be used for assessment of soil physical quality, this study has the hypothesis that the changes in the soil-water retention curve can change the references and responses of this index. Also, the changes in soil physical quality and carbon stocks under different weed management and cover crops in coffee plantations have not yet been investigated in Brazil.

Thus, our hypothesis is that weed control and the use of cover crops as a green manure between coffee rows changes the weed diversity and density, soil cover, soil carbon stocks and soil physical quality. Therefore, this chapter evaluates and provides information about the effects on soil total carbon stocks and soil physical quality caused by weed control and cover crops used as a green manure at different soil depths of a Latosol in a coffee plantation, in comparison to the soil under native forest.

2. Study site description and characterization

Since 2008, weed control and cover crop experiments have been conducted at the Agronomic Institute of Paraná (IAPAR) Experimental Station Farm in Londrina County, State of Paraná, Brazil (Latitude 23º 21' 30" S; Longitude 51º 10' 17" W of Greenwich).

The climate is Cfa – humid subtropical, according to Köppen's classification. The average temperature in the coldest month is lower than 18 ºC (mesothermal) and in the hottest months is higher than 22 ºC, creating a hot summer, with a low frequency of frost and a tendency to be rainy in the summer months, although without a dry season [14].

According to geomorphological mapping for the State of Paraná [23], Londrina is located in the morfoescultural unit Sedimentary Paraná Basin, morfoescultural units Third Plateau and morfoescultural sub-units Londrina Plateau.

The soil in the experimental area is derived from basalt and is classified as a Typical Dystroferric Red Latosol according to the Brazilian Soil Classification System [24]; Typic Haplorthox according to USDA soil taxonomy [25] and Ferralsol [26]. The slope of the study site is nearly level at 3%, and altitude is 550m above sea level.

The natural forest is amongst secondary mixed hardwood forest, close to the experimental area and the soil there provides a benchmark for soil quality. Some of its physical properties are shown in Table 1.

The soil particle-size distribution was determined by the pipette method [27], by chemical dispersion with a 5 mL 1 N sodium hydroxide solution in contact with the samples for 24 hours. Physical dispersion was accomplished by 2 hours, in a reciprocating shaker, which shakes 180 times per minute in a 38mm amplitude.

Field capacity and permanent wilting point were measured in the laboratory and corresponds to water contents remaining at the soil samples after saturation and equilibrated to matric potential - 33 kPa and - 1500 kPa, respectively, in a ceramic plate inside a pressure chamber.

Soil particle density was determined using a volumetric flask [28]. Total porosity was calculated by the soil bulk density to particle density ratio [29, 30].

Depth	Clay	Silt	Sand	SOC	FC	PWP	Bd	Pd	TP
cm	——— g kg^{-1} ———		g dm-3		—— cm^3 cm^{-3} ——		—— kg m^{-3} —		cm^{3}cm^{-3}
2 – 7	780	160	60	29.98	0.35	0.29	0.91	2.78	0.67
12 – 17	800	140	60	19.44	0.42	0.36	1.00	2.79	0.64
22 – 27	810	140	50	18.41	0.42	0.36	1.08	2.81	0.61
32 – 37	810	140	50	15.36	0.43	0.37	1.13	2.82	0.60

Depth: depth of sampling; SOC: soil organic carbon; FC: field capacity; PWP: permanent wilting point; Bd: bulk density; Pd: particle density and TP: total porosity. Averages from four replicates.

Table 1. Physical properties and total organic carbon content of a Typical Dystropherric Red Latosol under native forest adjacent to the study area at IAPAR in Londrina, North of State of Paraná.

The clay fraction dominated in all depths of this Dystroferric Red Latosol. The soil contained 250 – 280 g kg^{-1} of iron extractable by citrate-dithionite-bicarbonate, with hematite as dominant iron oxide, 620 — 650g kg^{-1} kaolinite and 20 — 40 g kg^{-1} Al-interlayered vermiculite [10].

The soil had a homogeneous structure, low soil bulk density, high total porosity and macroporosity and exhibited a granular structure like coffee powder throughout the profile, as described early by Kemper & Derpsch [31].

The coffee plantation was established about 30 years ago and the soil management history of the site included a conventional tillage system, the primary operation was disk ploughing (approximately 25cm soil depths) and the secondary was two disks acting to 15cm. Between 1978 to 2007 the weed control between coffee rows was done with disk harrowing and hand hoeing.

3. Experimental design, weed control and cover crops

The experimental area has been planted with Mundo Novo plants, spaced 3.50m between rows and 2.00m between plants, since 1978.

In 2008, cover crops and weed management systems were established in a randomized complete block design with four replicates. Each plot has two inter-rows and has an area of 112m^2 (7m x 16m) for each treatment (28 plots in total). The experimental design further included a split-plot, with each weed control and cover crop in the inter-rows as the main-plot factor and the soil sampling depths (2 – 7 cm, 12 – 17 cm, 22 – 27 cm and 32 – 37 cm) as a split-plot.

The weed and cover crops management systems (TREATMENTS) were as follows:

1. hand weeding (HAWE): performed with the aid of a hoe, when the weed reached 45 cm height, between August 2010 and July 2011 was performed four times.

2. portable mechanical mower (PMOW): with the aid of a portable knapsack mechanical mower

3. pre-emergence and post-emergence herbicides (HERB): A) pre-emergence: oxyfluorfen at a rate 4.0 L ha^{-1} of commercial product at 240 g L^{-1} (0.96 Kg active ingredient ha^{-1}), applied three times since beginning of the experiment, in November 2008, October 2010 and September 2011; B) post- emergence herbicides: glyphosate, at a rate 4.0 L ha^{-1} of commercial product at 360 g L^{-1} (1.44 Kg active ingredient ha^{-1}) applied six times (January, April, October and December 2009, April and December 2010); in March 2011 carfentrazone-ethyl was used as post-emergence herbicide at a rate 100 m L ha^{-1} of commercial product at 400 g L^{-1} (0.04 Kg active ingredient ha^{-1}).

4. cover crop peanut horse (*Arachis hypogeae*) used as a green manure (GMAY): was sown annually on October 23/2009; 14/2010 and 27/2011.

5. dwarf mucuna (*Mucuna deeringiana*) (Bort.) Merr used as a green manure (GMMD): was sown annually in October 23/2009; 14/2010 and 27/2011.

6. no-weed control between coffee rows (NWCB): the weed plants were left to grow freely between coffee rows.

7. no-weed control between coffee rows or under canopy of the coffee plants (weed check -WCCK).

8. native forest (NAFT): adjacent to coffee cultivation is a secondary mixed hardwood forest, located about 500m from experimental area.

Between each coffee row, two rows of the cover crops were sown annually at the beginning of the spring in October (23/2009; 14/2010 and 27/2011) and cut at flowering stage within the production cycle of the coffee.

It was observed in the field, that the species *Mucuna deeringiana* (Bort.) Merr grew faster than *Arachis hypogeae* until the end of December (sowing to flowering), after this stage the soil covered by these two species was similar.

4. Soil sampling and analyses

The soil sampling and analyses were performed in 2011 (the third year of this experiment) to assess the effects of weed control and cover crops on soil structure. The undisturbed soil samples were collected from the centre of the inter-rows between coffee plants (1.75m from the coffee stem) using a mechanical extractor and inox rings, 5cm high and 5cm in diameter. Also, to calculate the total carbon stocks, disturbed soil samples were collected under the coffee canopy at the same depths.

As reported previously, the undisturbed soil samples were collected at depths 2 – 7 cm, 12 – 17 cm, 22 – 27 cm and 32 – 37 cm. These depths were chosen for sampling because the surface layers are more relevant when assessing the impact of management on carbon stocks and are more frequently modified directly by cultivation [16]. These authors showed that the layers between 0 and 18 cm were most influenced by management. Furthermore, these layers are more influenced by weed control in coffee plantations, as shown earlier by Alcântara & Ferreira [9] and Araujo-Junior et al. [6, 7]. The selection of the fixed sampling depth, as done in this study, is somewhat arbitrary, but it must be identical for all profiles being compared and include the soil layer most susceptible to the influence of management [16].

The photos 1A and 1B show the no-weed control between coffee rows and dwarf mucuna used as a green manure and cover crop. Cover crops provide a good soil cover and protect the soil against hydric erosion and soil surface crusting. The soil cover with peanut horse (Figure 1C) and weed control with herbicides provided lower soil cover (Figure 1D).

Photo 1. Experimental plots: (A) weed check no-weed control between coffee rows and under coffee canopy, (B) plants of dwarf mucuna (*Mucuna deeringiana*) used as a cover crop and green-manure, (C) peanut horse (*Arachis hypogeae*) and (D) herbicides.

5. Total soil organic carbon and carbon stocks

Total soil organic carbon was determined by wet digestion following organic oxidation by $Cr_2O^{2-}_7$ in acid [32]. The total soil organic carbon concentrations in kg Mg^{-1} were obtained directly from chemical analyses for the two sites of sampling (under the coffee canopy and between coffee rows). Total soil organic carbon masses in each soil layer in Mg ha^{-1} were calculated from the thickness of the soil layer (0.10m) and the average soil bulk density in each layer, according to Equation 1, proposed by Ellert & Bettany [16].

$$M_{COT} = conc \cdot \rho_b \cdot T \cdot 10\,000\ m^2ha^{-1} \cdot 0.001\ Mg\ kg^{-1} \tag{1}$$

Where, M_{COT} total soil organic carbon mass per unit area (Mg ha^{-1}), conc is total soil organic carbon concentration (kg Mg^{-1}), ρb is the soil bulk density (Mg m^{-3}) and T thickness of soil layer (m).

6. The soil-water retention curve and its properties

Evaluation of soil physical quality includes measurements of the soil-water retention data and its properties performed in quadruplicate. Undisturbed soil samples were prepared for the exact size of inox rings. These soil samples were saturated with water for 48 hours. After that, undisturbed soil samples were equilibrated to a matric potential expressed as pressure head h (cm) of 20cm, 40cm, 60cm and 100cm on a suction table [33] (Eijkelkamp Equipment, P.O Box 4, 6987 ZG Giesbeek Nijverheisdsstraat 30, 6987 EM Giesbeek) and 330cm, 1,000cm, 5,000cm and 15,000cm in a ceramic plate inside a pressure chamber (Soil Moisture Equipment Crop., P.O. Box 30025 Santa Barbara, CA 93105) [34].

To calculate soil bulk density, undisturbed soil samples were dried in the oven at 105–110 °C for 48 hours to determine dry soil weight per unit volume [35, 36]. The volumetric soil water content was estimated using gravimetric soil water content times soil bulk density [37].

The soil microporosity was determined for the soil samples equilibrated to a matric potential - 6 kPa in a suction table, which separated the pores with effective diameter greater than 50 μm, drained from the cores (macropores). The soil macroporosity was calculated by the difference among total porosity and microporosity, which corresponds to water drained between 0 to 60 cm pressure head.

The soil-water retention curve is the functional relationship between water pressure head (cm) or soil matric potential (Ψ) versus soil water content (cm^3 cm^{-3}) was obtained for each undisturbed soil sample. The soil-water retention was fitted through the van Genuchten [17] model with Mualen [38] constraint (m = 1-1/n) described by the Equation 2, using software Soil Water Retention Curve (SWRC) [39].

$$\theta = \theta_{res} + \frac{\theta_{sat} + \theta_{res}}{\left[1 + (\alpha\,\Psi)\,n\,\right]^{1-1/n}} \tag{2}$$

Where, θ, θ_{res} and θ_{sat} represent the volumetric, residual and saturated soil water contents ($cm^3\ cm^{-3}$), respectively; α, m and n are the parameters of the fitted model that are related to scaling factor and the shape of the fitted curve; Ψ is the pressure head (cm).

The angular coefficient of the soil-water retention curve at inflection point (soil physical quality [S index]) was calculated by Equation 3 [22].

$$S_{index} = -n\,(\theta_{sat} - \theta_{res})\left[\frac{2n - 1}{n - 1}\right]^{-\left(\frac{1}{n} - 2\right)} + \theta_{res} \tag{3}$$

Data for soil cover, soil bulk density and total soil organic carbon were submitted to analysis of variance through the software SISVAR [41], considering a split plot design, comparing different weed management and cover crops in each soil depth. Linear regressions were obtained for soil macroporosity and soil physical quality S-index to obtain the lower boundary limit for this index.

7. Results and discussion

The results supported that in Oxisol planted with coffee plantation the weed diversity and density, soil cover, soil carbon stocks and soil physical quality measured by S index and macroporosity are related to weed control and cover crops.

Seven weed species were identified in the coffee plantation and were submitted to different weed control and cover crops between coffee rows and under the coffee canopy in May 2011 (Table 2). Although the soil in the present study has a clayey texture, the number of weed species was relatively small compared to the previous study of a coffee plantation assessed in the summer season in a tropical region (December 2007) [42].

Carter & Ivany [43] highlighted that the soil type and kind of tillage can significantly influence weed seed bank composition. They also explained that reduced physical protection and aggregate entrapment in sandy, compared to clay textured soils, would allow weed seeds to move to deeper soils depths (12cm), where seed dormancy would be independent of soil texture. In addition to this, Carter & Ivany [43] apud Albrecht and Pilgram showed that soil textures are mainly related to soil-water retention and can significantly influence weed seed density, weed composition and seed size, through selective pressure on available water capacity.

The low density and diversity of weed species observed in this study (Table 2) was probably due to low temperature, with a mean monthly temperature of 17.7ºC and mean precipitation of 93mm. However, in May 2011 the rain distribution was erratic, with only 7.6mm of precipitation, which impaired the weed germination, growth and development.

A greater diversity and density of weed plants was observed in the soil between coffee rows, compared to under the canopy (Table 2). This suggests that the coffee canopy promoted weed suppression.

SCIENTIFIC NAME AND WEED SPECIE	Weed management and cover crops in a coffee plantation													
	HAWE		PMOW		HERB		GMAY		GMMD		NWCB		WCCK	
	BCR	UCC	BCR	UCC	BCR	UCC	BCR	UCC	BCR	UCC	BCR	UCC	BCR	UCC
Portulaca oleracea L. Purslane							X		X					
Digitaria insularis (L.) Fedde Sourgrass	X		X		X						X		X	
Brachiaria decumbens Signal grass											X		X	
Sida rhombifolia Arrowleaf sida													X	
Talinum paniculatum (Jacq.) Gareth. Fameflower		X		X	X	X	X	X	X	X			X	X
Momordica charantia L. Bitter melon			X	X							X			X
Phyllanthus tenellus Roxb. Leafflower	X	X	X	X	X	X	X	X	X	X	X			X

HAWE: hand weeding; HERB: pre plus post-emergence herbicides; PMOW: portable mechanical mower; GMAY: cover crop peanut horse *Arachis hypogea*; GMMA: cover crop dwarf mucuna *Mucuna deeringiana*; NWCB: no-weed control between coffee rows; WCCK: weed check. BCR: between coffee rows; UCC: under coffee canopy.

Table 2. Weed species distribution under different management and cover crops in a coffee plantation at two positions, between coffee rows and under the coffee canopy in May 2011.

Among the management adopted, a greater density and diversity of weed plants (Table 2) was detected in the no-weed control between coffee rows and weed check. The absence of soil disturbance in these weed management systems allows formation of a bigger and more diverse weed seed bank in soil [43, 44, 7]. However, the former authors suggested the diversity is not directly related to higher infestation levels.

The species fameflower and leafflower were observed in almost all treatments, except in the no-weed control between coffee rows (Table 2).

8. Soil cover

The weed control and cover crops had a significant effect on soil cover values offered by weed plants between coffee rows and under the coffee canopy (Figure 1). For the assessment in May 2011, soil cover by weed plants and cover crops between coffee rows was in the following order: NWCB = WCCK > HAWE = PMOW > GMAY = GMMA > HERB. However, it should be noted that managements HAWE and PMOW also promoted a good soil cover at this time of year.

Between coffee rows, no-weed control and weed check (without weed control between coffee rows and under canopy) were most effective in soil protection, whereas the weed control with herbicides was less effective in soil protection. Intermediate levels of effectiveness were observed for the hand weeding, portable mechanical mower and green manures.

Under the coffee canopy, significant differences were not evident among most of the managements. Greater soil cover was obtained by hand weeding, dwarf mucuna, no-weed control between coffee rows and weed check (without weed control between coffee rows and under canopy) (Figure 1).

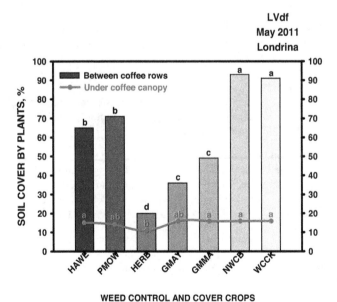

Figure 1. Soil cover by weed and cover crops in a coffee plantation between coffee rows and under coffee canopy in May 2011. HAWE: hand weeding; HERB: pre plus post-emergence herbicides; PMOW: portable mechanical mower; GMAY: cover crop peanut horse (*Arachis hypogea*); GMMA: cover crop dwarf mucuna (*Mucuna deeringiana*); NWCB: no-weed control between coffee rows; WCCK: weed check.

These results show the potential that weed plants have to provide protection between coffee rows against the direct impact of raindrops, thus reducing the potential for loss of water and soil. Despite the high water infiltration rate of Dystropherric Red Latosol in Londrina (70mm h^{-1}), there is an intense rainfall erosivity index, with over 1,000 MJ.mm.ha^{-1}.year^{-1} [2, 15]. Therefore, this soil can experience losses exceeding 100 tons/ha on bare soil and 33 tons/ha where crops are grown [2]. These authors showed that the squaring operation (Post-harvest coffee) is critical to soil losses in coffee crops grown, due to removal of lower leaves and soil protection. Thus, weeds and cover crops culti-vated between coffee rows might be used to protect the soil against direct raindrop im-pact and reduce water and soil losses in coffee plantations.

9. Soil bulk density

The soil samples from the coffee-cultivated plots subjected to different weed manage-ment and cover crops between coffee rows had a higher bulk density at four depths, when compared to the soil samples from native forest soil (Table 3). Some previous stud-ies have shown this increase in the soil bulk density under coffee plantation in relation to native forest [7, 9].

The soil bulk density for Latosol at 2—7 cm, 22—27 cm and 32—37 cm depths were not sig-nificantly varied between different weed and cover crops management. However, at 12—17 cm depths obvious differences of soil bulk density were detected. The soil bulk density for Latosol under HAWE, NWCB and WCCK weed managements and both cover crops used as green manure, resulted in higher packing of solid particles of the soil (Table 3). The higher bulk densities at 12—17 cm depth might be due the stress concentration applied by tyres and equipment used for weed control in the past, which promoted a higher degree of physi-cal degradation and packing of solid particles of the soil.

Nevertheless, neither soil bulk densities found in the present study were considered higher than the critical soil bulk density (1.20kg dm^{-3}) for coffee root growth established by Araujo-Junior et al. [7] in Latosol with 560g kg^{-1} clay. These results are in agreement with Streck et al. [45], who obtained the critical soil bulk density, based on soil physical quality S-index, for seven Latosols under different land uses with clay content between 160 to 760 g kg^{-1}.

For the soil in this study, Derpsch et al. [46] suggested the value equal to 1.20 kg dm^{-3} for problems with root growth and soil aeration are not probable. On the other hand, ac-cording to these authors, values of soil bulk density higher than 1.25 kg dm^{-3} might re-strict root growth.

In a Dystroferric Red Latosol with 800g kg^{-1} clay, Tormena [47] observed that soil physical quality measured by S-index decreased as soil bulk density or compaction increased as a re-sult of reducing macropores volume, with a consequent alteration on the pore size distribu-tion. They found that at 1.16kg dm^{-3} there are restrictions on soil physical quality associated with soil resistance to root penetration. However, they pointed out that using S-index instead of soil bulk density values has the advantage of getting similar S values in soils of different particle size distribution.

Native forest / weed and cover crops	Soil bulk density, kg dm⁻³			
	Depths, centimetres			
	2 – 7	12 – 17	22 – 27	32 – 37
Native forest	0.91 A	1.00 A	1.08 A	1.13 A
Hand weeding	1.10 B	1.16 C	1.16 B	1.14 A
Portable mechanical mower	1.11 B	1.11 B	1.12 B	1.15 A
Herbicides	1.11 B	1.10 B	1.19 B	1.18 A
Peanut horse Arachis hypogaea	1.12 B	1.23 C	1.18 B	1.13 A
Dwarf mucuna Mucuna deeringiana	1.12 B	1.19 C	1.21 B	1.19 A
No-weed control between coffee	1.08 B	1.18 C	1.14 B	1.09 A
Weed check	1.07 B	1.20 C	1.17 B	1.15 A

Averages followed by the same uppercase letters compare different treatments in each soil depth.

Table 3. Soil bulk density for Latosol samples collected between coffee rows in four depths under different weed control and cover crops managements.

For seven Oxisols in the South of Brazil under different land uses Streck et al. [45] showed lower values for soil bulk density under native forest than for the soil under direct drilling. They showed no relationship between clay content and dispersible clay in water using soil physical quality S-index. However, they found an exponential decay relationship between S-index vs. soil bulk density and S-index vs. precompression stress. Although Tormena et al. [47] did not comment on the kind of relation between S-index and soil bulk density, it was possible to note similar behaviour to exponential decay.

10. Total soil carbon stocks

Figure 2 shows the total soil organic carbon stocks for a Dystroferric Red Latosol at four depths, under natural forest and coffee plantation, submitted to different weed controls and cover crops.

The soil under native forest contained lower carbon stocks compared to the soil under coffee plantation submitted to different weed management systems. This is likely due to the lower soil mass per unit area under native forest and also the large amount of weed dry mass added during thirty years, resulting in soil organic carbon accumulation in the soil under the coffee plantation (Figure 2).

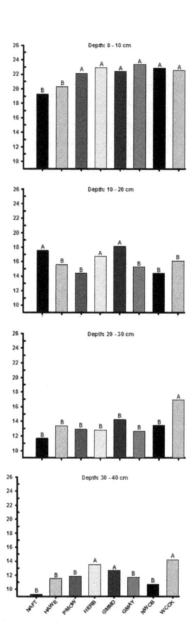

Figure 2. Total soil organic carbon stocks for a Dystroferric Red Latosol at four depths, under natural forest and coffee plantation. NATF: native forest; HAWE: hand weeding; HERB: pre plus post-emergence herbicides; PMOW: portable mechanical mower; GMMD: dwarf mucuna (*Mucuna deeringiana*); GMAY: peanut horse (*Arachis hypogea*); NWCB: no-weed control between coffee rows; WCCK: weed check.

In coffee plantations, the planning of weed control, with the input of cover crops as a green manure, restored the carbon stocks to similar levels as the native forest. Another possible explanation for the higher stocks of carbon in the coffee plantation is physical protection of organic matter by aggregates and organomineral interactions [49] because this Typical Dystroferric Red Latosol (Typic Haplorthox) has a clay fraction that is dominated by hematite and phyllosilicate as kaolinite [10]. The organomineral interactions, such as iron and aluminium oxides and the surface functional groups of organic matter, interfere with the decomposition of organic matter by microbial processes, even under conventional tillage [49].

Based on information supported by Castro Filho et al. [48], in experiments of crop succession in conventional tillage and direct drilling over 14 years, Bayer et al. [49] estimated the total carbon stocks at 0—20 cm depths and the rate of carbon inputs for clayey Latosol from IAPAR. They found 27.40 to 29.00 Mg C stock ha^{-1} under conventional tillage and 31.87 to 32.30 Mg C stock ha^{-1} under direct drilling at a rate 0.24 to 0.48 Mg ha^{-1} year^{-1}.

At 0 — 20 cm depth, our results were: NAFT: 36.79 Mg C stock ha^{-1}; HAWE: 35.86 Mg C stock ha^{-1}; PMOW: 36.50 Mg C stock ha^{-1}; HERB: 39.65 Mg C stock ha^{-1}; GMAY: 37.63 Mg C stock ha^{-1}; GMMA: 40.57 Mg C stock ha^{-1}; NWCB: 37.28 Mg C stock ha^{-1} and WCCK: 38.64 Mg C stock ha^{-1}. Generally, the conversion of native forest into crops can promote losses of soil carbon stocks. However, this study showed that with integrated weed management and cover crops between coffee rows the carbon stocks can be maintained or increase.

The carbon stocks under coffee plantation were higher than those estimated by Bayer et al. [49] for the tillage treatments, based on results obtained by Castro Filho et al. [48]. These results might be due to the lower decomposition rates between coffee rows or greater biomass inputs from weed populations and cover crops. Also, the high coffee population density might contribute to lower soil temperature and increase in soil organic carbon in the coffee plantation, as reported earlier by Pavan et al. [3].

The total soil carbon stocks at 10 to 20 cm depths for NAFT were similar to HERB and GMMD. Calegari et al. [49], in a long-term experiment supported by results for a clayey Rhodic Hapludox with 720 g kg^{-1} clay, from Pato Branco (Southwestern of State of Paraná), found that the weed provided some increase in soil organic carbon compared to burning. Also, they observed the effects of several winter crops and tillage treatments over 19 years. They found 68.86 Mg C ha^{-1} under no-tillage and 65.21 Mg C ha^{-1} under conventional tillage between 0 to 20 cm depths. Another important result found by these authors, was that independent of soil tillage, the total soil organic carbon stocks decreased in the following order: lupin > oat > radish > vetch > wheat. Although lupin was intermediate in dry mass production compared to others. These results highlight that winter cover crops help increase the soil carbon stocks compared to wheat [50].

Early reports from experiments done at IAPAR, between 1964 and 1978, showed that organic matter content decreased by approximately 45% through coffee cultivation, compared to native forest [31]. However, in that time, the coffee plantations had mechanized weed control by tillage, which increased soil losses by removing organic substances and nutrients.

However, current uses of integrated weed management systems and cover crops between coffee rows can promote higher organic matter accumulation on the soil surface, increasing protection against soil erosion and nutrients losses.

The use of integrated weed management systems [5, 7, 9], coffee population density [3] and cover crops have been suggested to play an important role in soil carbon stocks [47 to 49]. The results found in this study, suggest that integrated weed management and cover crops between coffee rows helps the maintenance of soil carbon stocks.

It was observed that the cover crop peanut horse provides good carbon accumulation through the root system. In comparison, assessments carried out in 2010 and 2011 suggest the values of the total soil organic carbon under the cover crop peanut horse increased by 4 g dm^{-3} at 0 – 10 cm depth (unpublished data).

11. Soil physical quality

The soil-water retention curve for the Genuchten-Mualem equation for the Latosol submitted to different weed control and cover crops in a coffee plantation at four depths was significant at a 1% probability level, for t-Student test. The coefficient of determination (R^2) ranged from 0.71 to 0.99.

The residual soil-water content (θ_{res}) ranged from 0.26 cm^3 cm^{-3} for the samples collected from WCCK at 2 — 7 cm depth, to 0.40 cm^3 cm^{-3} for the HERB at 22 — 27 cm depth. Based on this information, it was possible to see that weed control with HERB increased soil water retention at high-pressure heads (15,000 cm col H$_2$O) and this management promoted close pore-size distribution. On the other hand, the WCCK promoted lower water retention, which indicates high pore diameter distribution on soil samples under this management.

The saturated soil-water content (θ_{sat}) ranged from 0.68 cm^3 cm^{-3} for the soil samples collected from NAFT at 2 — 7 cm depth, to 0.52 cm^3 cm^{-3} for the HERB at 22 — 27 cm depth. These results suggest higher total porosity of soil samples from NAFT and lower in HERB.

The value of α ranged from 0.0292 to 0.8065 (1/cm) at which the retention curve becomes the steepest, as reported earlier by van-Genuchten [17]. The value of the parameter "n" ranged from 1.2386 for NAFT land use at 12 — 17 cm depth to 1.8321 for HAWE weed control. The smaller value of n represents a less steep soil-water retention curve [17] and "m" from 0.1927 for the soil samples under NAFT at 12 — 17 cm depth to 0.4338 for the samples under HERB (22 — 27 cm).

In general, the soil physical quality of samples from Latosol quantified by S-index under coffee plantation and native forest at four depths (Figure 3) were higher than the lower boundary limit established by the regressions based on macroporosity 0.10 cm^3 cm^{-3} considered as critical for soil aeration (Figure 4).

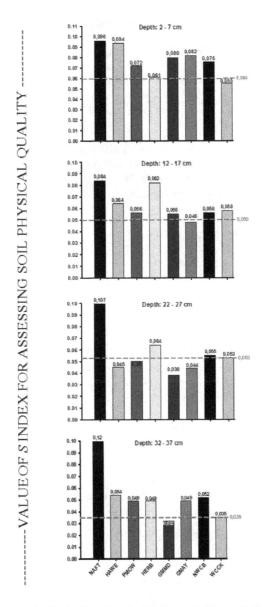

Figure 3. Soil physical quality "S index" for a Dystroferric Red Latosol in 2–7 cm, 12–17 cm, 22-27 cm and 32–37 cm layers, under natural forest and coffee plantation. NATF: native forest; HAWE: hand weeding; HERB: pre plus post-emergence herbicides; PMOW: portable mechanical mower; GMMD: dwarf mucuna (*Mucuna deeringiana*); GMAY: peanut horse (*Arachis hypogea*); NWCB: no-weed control between coffee rows; WCCK: weed check. The dotted horizontal line represents the critical value for *S index* in each soil layer.

It should be noted that values of S-index obtained in the present study are higher than the reference values suggested by Dexter [22] and found by many studies for Brazil's tropical soils [47, 51, 52]. In an overview of the relationship between S-index and soil physical properties (particle size distribution, bulk density, total porosity, macroporosity) from 2,364 soil samples with a wide range of clay content, Andrade & Stone [53] found that lower boundary limit for S-index is equal to 0.045. This proved to be adequate to separate soils with good structure and soils with a tendency to have poor soil structure, where values of S ≤ 0.025 indicate physically degraded soils.

Dexter [22] suggested that the boundary between soils with good and poor soil structural quality occurs at values of approximately S = 0.035. Values of S < 0.020 are clearly associated with very poor soil physical quality in the field. Though, in this study we fitted soil-water retention curves using volumetric soil water content, which promoted higher S values, in agreement with Maia [52].

In the present study, the soil-water retention curve was adjusted for volumetric water content to improve the response of S-index to soil compaction. Under soil compaction, there are changes in volumetric water content and there is no change in gravimetric water content, which can improve the sensitivity of S-index.

Dexter [22] and Maia [52] suggested that the soil-water retention curve must be fitted by gravimetric soil water content to use the reference values established by the former. Although it could equally be defined using the volumetric water content, changing reference values for assessing soil physical quality as suggested in the present study.

In all depths, the highest soil physical quality S-index was observed for soil under natural forest (Figure 3), which is due to the absence of stress history, which was observed by high macroporosity and lowest soil bulk density (Table 3). These results highlight that although the adoption of weed control without machine traffic and cover crops as a green manure between coffee rows ameliorates slightly those harsh effects on soil quality, the impacts of the coffee plantation in relation to the soil under native forest are highly significant.

The soil physical quality quantified by S-index at 2 — 7 cm depths in plots under hand weeding was similar to the value observed in the soil under native forest (Figure 3). This result might be due the effect of the hoe, which loosens the soil surface, promoting an increase in the total porosity and a decrease in soil bulk density (Table 3).

After three years studying a Rhodic Paleudalf (Nitossolo Vermelho distroférrico) with 600g kg⁻¹ clay under crop rotation and chiselling, Calonego & Roslem [51] observed a higher S-index value as a result of better soil management compared to the beginning of their experiment. They observed mainly soil physical quality improvements on the soil surface, due to the chiselling and loosening of the soil and also as a result of greater root growth in this soil layer.

At 22 — 27 and 32 — 37 cm depths, soil physical quality S-index in plots under dwarf mucuna (*Mucuna deeringiana*) (Bort.) Merr had lower values than the critical limit (Figure 3). These findings must be due to the stress history caused by the use of the mechanical rotary tiller and disk harrowing as part of weed control between coffee rows in the past, before the experiment installation. In clayey soil in Northern Paraná, the excessive use of heavy plough-ing harrow equipment compacted the subsurface layers, accelerated erosion, decreased infiltration rate, inhibited root development and reduced crop productivity [31, 56].

Similar results were obtained by Calonego & Rosolem [51], mainly in the 27.5 to 32.5 cm lay-er under triticale plus pearl millet. This characterizes a soil with poor structural quality, with the lowest S = 0.019. They suggested that crop rotation involving only monocotyledonous species, limited the cultivation effect on the soil structure to the first 20cm of the soil depth. Although some cover crops have appeared to improve soil protection against erosion and compaction, improve water infiltration rate, soil-water retention and soil carbon stocks, some of them did not show a beneficial effect at deeper soil layers, since the root system is concentrated at the soil surface.

In a Cerrado Red Latosol with 420 g kg^{-1} clay under direct drilling over four years Silva et al. [55] observed that the sills of the active parts of the disk plough and disk harrow increased soil strength and reduced the saturated hydraulic conductivity in layer below the sills of this equipment.

In deeper soil layers (22 — 27 cm and 32 — 37 cm) the differences among the S-index calcu-lated for the Latosol samples under native forest and for coffee plantation were greater (Fig-ure 3), which suggests that these depths had lower soil physical quality. These results might be due to the lower organic carbon content in coffee plantation, which favours closer pack-ing of solid soil particles, as a result of decreased macroporosity and increased soil bulk den-sity, [7] and consequently decreases soil physical quality index in comparison to soil without stress history (native forest). Furthermore, the organic matter content reflects the degree of soil degradation in clayey soil derived from basalt rocks [56]. A decrease in organic matter content over the time reflecting the inadequate land use was observed.

In the past, measurements of the same experimental field have shown that the reduction of soil organic matter due to tillage operations can contribute to the destruction of natural po-rosity and create a compact layer in clay soils in the North of the State of Paraná [31]. On the other hand, in surface layers, the weed and cover residues are left as mulch, so the differen-ces in soil physical quality index were less marked compared the soil under native forest.

12. Relationship between S-index and soil macroporosity

Figure 4 shows the relationship between the soil macroporosity and S-index for the soil un-der different weed management, cover crops in a coffee plantation and in soil under native forest. For all depths, the S-index increased linearly with increasing soil macroporosity (Fig-ure 4). These results are in agreement with Andrade & Stone [53] who observed that S-index increased with total porosity and macroporosity.

The regression lines fitted to all the data in Figure 4 explained 70% to 88% of the variance in S-in-dex. All regressions for the Dystropherric Red Latosol were significant at 1% probability level, by t-Student test. Based on these equations, the S-index of the soil surface (2 — 7 cm depth) increased less as macroporosity increased. On the other hand, at 32 — 37 cm depth the soil physical quality quantified through S-index increased in greater proportion with soil macroporosity.

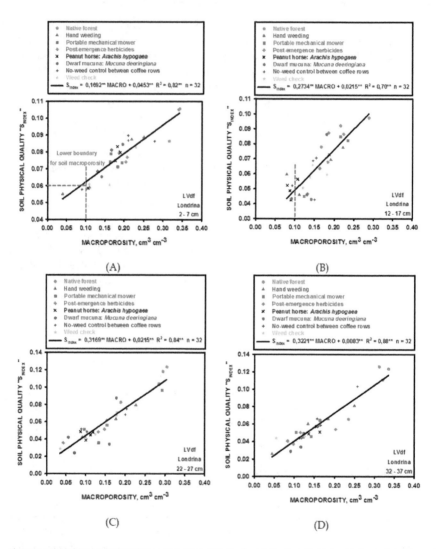

Figure 4. Relationship between soil macroporosity and S-index for a Dystroferric Red Latosol in four soil depths, under natural forest and coffee plantation. The lower boundary for soil macroporosity considered was 0.10 cm³ cm⁻³ (A) soil samples taken from 2 – 7 cm depth; (B) 12 – 17 cm depth; (C) 22 – 27 cm depth and (D) 32 – 37 cm depth.

The regressions lines in Fig. 4 A to 4 D were used to define the lower boundary for soil physical quality for Latosol cultivated with coffee plantation and might be used for predicting soil physical quality (S-index) through macroporosity of this Latosol under different weed control and cover crops. With inadequate soil management, a flattened soil-water retention curve was observed, with a reduction in the slope of this curve at the inflection point [21, 20, 42]. Thus, it can be inferred that soil compaction changes pore diameter, but not all pores are reduced similarly [20, 21, 42, 45, 47, 51 – 54]. Typically, compacted Oxisols have low macroporosity and total porosity and, as a consequence, have low infiltration rate [31].

Many studies have shown that the macropores (pores with effective diameter greater than 50 μm) are reduced first under stress. Then, compaction has a great influence on macropore flow, but there have been few attempts to quantify these effects [20]. Han et al. [54] found that characteristics of pore diameter at the inflection point were related to the hydraulic conductivity. Due to that, we relate the S-index with macroporosity (Figure 4) and use this relation to define the lower boundary limit for S-index for four depths.

13. Conclusions

The results supported the hypothesis that weed control and cover crops between coffee rows change the weed diversity and density, soil cover, soil carbon stocks and soil physical quality measured by S-index and macroporosity. Also, the weed control and cover crops between coffee rows ameliorate slightly the harsh effects of the coffee crop system on total soil carbon stocks and soil physical quality in the North of the State of Paraná.

Adjustment of the soil-water retention curve changed the references and responses of the S-index. Based on S-index, it was observed that the soil under coffee plantation, submitted to different weed controls and cover crops as a green manure between coffee rows without traffic machines, contributed to preserve soil physical quality in soil depths between the surface and 40cm, except the plots under the cover crop dwarf mucuna at 22 — 27 cm and 32 — 37 cm soil depths. Therefore, the integration of weed management and cover crops must be recommended to help maintain carbon stocks and improve soil physical quality between coffee rows.

Acknowledgements

The authors are grateful to Technical Scientific Directors of Agronomic Institute of Paraná – IAPAR and to Brazilian Coffee Research and Development of Consortium – CBP&D-Café for the financial aid for execution of the work. The authors also would like to thank Dr. Élcio Libório Balota and MSc. Oswaldo Machineski for their assistance with the data under the coffee canopy.

Author details

Cezar Francisco Araujo-Junior[1], Benedito Noedi Rodrigues[2], Júlio César Dias Chaves[1] and George Mitsuo Yada Junior[3]

*Address all correspondence to: cezar_araujo@iapar.br

1 Soils Area at Agronomic Institute of Paraná – IAPAR, Londrina, State of Paraná, Brazil

2 Plant Production Area at Agronomic Institute of Paraná – IAPAR, Brazil

3 Scientific Initiation Program at Agronomic Institute of Paraná ProICI - IAPAR, Federal University Technology of Paraná – UTFPR, Londrina, PR, Brazil

References

[1] Lal R. Degradation and Resilience of Soils. Philosophical Transactions of the Royal Society, 1997, 352(5) 997–1010, ISSN 0962-8436.

[2] Rufino RL, Henklain JC, Biscaia RCM. Influência das Práticas de Manejo e Cobertura Vegetal do Cafeeiro nas Perdas de Solo. (In Portuguese, with English abstract). Revista Brasileira de Ciência do Solo, Brazilian Journal of Soil Science, 1985, 9(3) 277–280, ISSN 0100-0683.

[3] Pavan MA, Chaves JCD, Siqueira R, Androcioli Filho A, Colozzi Filho AA, Balota EL. High coffee population density to improve fertility of an Oxisol. Pesquisa Agropecuária Brasileira, 1999, 34(3), 459 — 465, ISSN 0100-204X.

[4] Carvalho R, Silva MLN, Avanzi JC, Curi N, Souza FS de. Erosão Hídrica em Latossolo Vermelho sob Diversos Sistemas de Manejo do Cafeeiro no Sul de Minas Gerais. (In Portuguese, with English abstract). Ciência & Agrotecnologia 2007, 31(6), 1679–1687, ISSN 1413-7054.

[5] Faria JC, Shaefer CER, Ruiz HA, Costa LM. Effects of Weed Control on Physical and Micropedological Properties of Brazilian Ultisol. Revista Brasileira de Ciência do Solo, 1998, 22(3) 731 — 741, ISSN 0100-0683.

[6] Araujo-Junior CF, Dias Junior M de S, Guimarães PTG, Pires BS. Resistência à compactação de um Latossolo cultivado com cafeeiro, sob diferentes sistemas de manejos de plantas invasoras. (In Portuguese, with English abstract). Revista Brasileira de Ciência do Solo, 2008, 32(1) 25 – 32, ISSN 0100-0683.

[7] Araujo-Junior CF, Dias Junior M de S, Guimarães PTG, Alcântara EN. Capacidade de suporte de carga e umidade crítica de um Latossolo induzida por diferentes manejos. (In Portuguese, with English abstract). Revista Brasileira de Ciência do Solo, 2011, 35(1), 115 – 131, ISSN 0100-0683.

[8] Kurachi SAH, Silveira GM. Compactação do solo em cafezal provocada por diferentes métodos de cultivo. (In Portuguese, with English abstract). Instituto Agronômico de Campinas, 1984, 1—28.

[9] Alcântara EM, Ferreira MM. Efeitos de métodos de controle de plantas daninhas na cultura do cafeeiro (Coffea arabica L.) sobre a qualidade física do solo. (In Portuguese, with English abstract). Revista Brasileira de Ciência do Solo, 2000, 24(4) 711–721, ISSN 1806-9657.

[10] Castro Filho C, Logan TJ. Liming effects on the stability and erodibility of some Brazilian Oxisols. Soil Science Society American of Journal, 1991, 55(5) 1407–1413, ISSN 0361-5995.

[11] Braga NR, Wutke EB, Ambrosano EJ, Bulisani EA. Atualização do boletim 200. Campinas: Instituto Agronômico, 2006.

[12] Chaves JC, Rodrigues BN, Fantin D. Manejo do solo visando o controle de ervas, proteção e melhoria no ambiente da lavoura cafeeira. In: Simposio de Pesquisa dos Cafés do Brasil, 7., 2011, Araxá. Anais... Araxá: Editora, 2011.

[13] Chaves JC, Gorreta RH, Demoner CA, Casanova Junior G, Fantin D. O amendoim cavalo (Arachis hypogea) como alternativa para o cultivo intercalar em lavoura cafeeira. Londrina: Iapar, 1997. 20p. (Techical Bulletin of the Agronomic Institute of Paraná)

[14] Instituto Agronômico do Paraná - IAPAR. Cartas climáticas do Estado do Paraná. Londrina: Iapar, 1994. 49p.

[15] Waltrick PC, Machado MAM, Oliveira D de, Grimm AM, Dieckow J. Erosividade de chuvas no estado do Paraná: atualização e influências dos eventos "El NIÑO" e "LA NIÑA". Curitiba: DSEA, 2011. 20p. il. – (SBCS-NEP. (Technical Bulletin of the State Nucleo of Soil Science).

[16] Ellert BH, Bettany JR. Calculation of organic matter and nutrients stored in soil under contrasting management regimes. Canadian Journal of Soil Science, 1995. 75(5) 529 – 538, ISSN 0008-4271

[17] van Genuchten MTh. A closed-form equation for predicting the hydraulic conductivity of unsaturated soils. Soil Science Society of America Journal, Madison 1980; 44(5) 892 – 898, ISSN 0361-5995

[18] Vanapalli SK, Fredlund DG, Pufahl DE. The influence of soil structure and stress history on the soil-water characteristics of a compacted till. Géotechnique 1999;49(2) 143 — 159.

[19] Kutilek M, Nielsen DR. Soil Hydrology. Catena Verlag: GeoEcology textbook; 1994.

[20] Alaoui A, Lipiec J, Gerke HH. A review of the changes in the soil pore system due to soil deformation: A hydrodynamic perspective. Soil & Tillage Research 2011; 115 — 116(1) 1 — 15.

[21] Jury WA. Advanced soil physics – Lectures Notes. University of California. 1979. 172 p.

[22] Dexter AR. Soil physical quality. Part I. Theory, effects of soil texture, density, and organic matter, and effects on root growth. Geoderma 2004; 120(1) 201 – 214, ISSN 0016-7061

[23] Santos L JC, Oka-Fiori C, Canali, N E, Fiori, A P, Silveira C T da, Silva J M F da, Ross J L S. Mapeamento geomorfológico do Estado do Paraná. (In Portuguese, with English abstract). Revista Brasileira de Geomorfologia, Brazilian Journal of Geomorphology, 2006; 7(2) 03–12.

[24] Empresa Brasileira de Pesquisa Agropecuária - EMBRAPA. Sistema brasileiro de classificação de solos. (2nd Ed.) Centro Nacional de Pesquisas de Solos. Rio de Janeiro: Embrapa Solos; 2006.

[25] Soil Survey Staff. Keys to soil taxonomy (8th ed), ISBN 2-853552-261-X. Washington: USDA-NRCS, 1998.

[26] Food and Agriculture Organization – FAO. World reference for soil resources: A framework for international classification, correlation and communication. ISBN, 92-5-105511-4, Rome: FAO; 2006.

[27] Day PR. Particle fractionation and particle-size analysis. In: Black, CA. (Ed.). Methods of soil analysis. Madison: American Society of Agronomy; 1965. n. 1, Part I. p545—567.

[28] Blake GR., Hartge KH. Partycle density. In: Klute A. (ed.). Methods of soil analysis. Part 1. 2 nd ed. Madisson: American Society of Agronomy; 1986b. p377—382.

[29] Vomocil JA. Porosity. In: Black CA. (ed.). Methods of soil analysis, Part1. Physical and Mineralogical Properties. Madison: American Society of Agronomy, 1965. p299—314.

[30] Flint LE, Flint AL. Porosity. In: Dane JH, Topp GC. (ed.) Methods of soil analysis: physical methods. Madison: Soil Science Society of America; 2002. p241–254.

[31] Kemper B, Derpsch R. Soil compaction and root growth in Parana, In: Russell RS, Igue K, Mehta YR. (ed.) Proceeding of the symposium on the soil/root system. Londrina: Instituto Agronômico do Paraná – IAPAR; 1980. p81–101.

[32] Walkley A, Black, IA. An examination of the Degtjareff method for determining soil organic matter and a proposed modification of the chromic acid titration method. Soil Science 1934; 37, 29–38.

[33] Romano N, Hopmans JW, Dane JH. Suction table. In: Dane JH, Topp GC. (ed.) Methods of soil analysis: physical methods. Madison: Soil Science Society of America; 2002. p692–698.

[34] Dane JH, Hopmans JW. Hanging water column. In: Dane JH, Topp GC. (ed.) Methods of soil analysis: physical methods. Madison: Soil Science Society of America; 2002. p680–683.

[35] Blake GR., Hartge KH. Bulk density. In: Klute A. (ed.). Methods of soil analysis. Part 1. 2 nd ed. Madisson: American Society of Agronomy; 1986a. p363—375.

[36] Grossman RB, Reinsch TG. Bulk density and linear extensibility. In: Dane JH, Topp GC. (ed.) Methods of soil analysis: physical methods. Madison: Soil Science Society of America; 2002. p201–228.

[37] Topp GC, Ferré PA. Water content. In: Dane JH, Topp GC. (ed.) Methods of soil analysis: physical methods. Madison: Soil Science Society of America; 2002. p417–424.

[38] Mualem Y. A new model for predicting the hydraulic conductivity of unsaturated porous media. Water Resources Research 1976;12(3)513—522.

[39] Dourado-Neto D, Nielsen DR, Hopman JW, Reichardt K, Bacchi OOS, Lopes PP. Soil Water Retention Curve (SWRC). Version 3.0, Piracicaba, 2001. Software.

[40] Dexter AR, Bird NRA. Methods for predicting the optimum and the range of water contents for tillage based on the water retention curve. Soil & Tillage Research 2001; 57(4) 203 — 212.

[41] Ferreira DF. Análises estatísticas por meio do SISVAR para Windows 4. 0. In: Reunião Anual da Região Brasileira da Sociedade Internacional de Biometria, 2000, São Carlos. Brazil. p. 255-258.

[42] Araujo-Junior CF, Dias Junior M de S, Guimarães PTG, Alcântara EN. Sistema poroso e capacidade de retenção de água de um Latossolo submetido a diferentes manejos de plantas invasoras em uma lavoura cafeeira. (In Portuguese, with English abstract). Planta Daninha 2011;29(3) 499 — 513.

[43] Carter MR, Ivany JA. Weed seed bank composition under three long-term tillage regimes on a fine sandy loam in Atlantic Canada. Soil & Tillage Research 2005;90(1) 29 — 38.

[44] Correia NM, Durigan JC. Emergência de plantas daninhas em solo coberto com palha de cana-de-açucar. Planta Daninha 2004;22(1) 11 — 17.

[45] Streck CA, Reinert DJ, Reichert JM, Horn H. Relações do parâmetro S para algumas propriedades físicas de solos do sul do Brasil. Revista Brasileira de Ciência do Solo, 2008;32(Especial) 2603 — 2612, ISSN 0100-0683.

[46] Derpsch R, Roth CH,. Sidiras N & Köpke U. Controle da erosão no Paraná, Brasil: sistemas de cobertura do solo, plantio direto e preparo conservacionista do solo. Fundação Instituto Agronômico do Paraná e Deutsche Gesellschaft für Technische Zusammenarbeit (GTZ) GmbH. Eschborn; 1991. ISBN 3-88085-433-5.

[47] Tormena CA, Silva AP, Imhoff SCD, Dexter AR. Quantification of the soil physical quality of a tropical Oxisol using the S index. Scientia Agricola 2008; 65(1) 56 — 60.

[48] Castro Filho C, Muzilli O, Podanoschi AL. Estabilidade dos agregados e sua relação com o teor de carbono orgânico num Latossolo Roxo distrófico, em função de sistemas de plantio, rotação de culturas e métodos de preparo das amostras. Revista Brasileira de Ciência do Solo, 1998;22(3) 527 — 538, ISSN 0100-0683.

[49] Bayer C, Martin-Neto L, Mielniczuk J, Pavinato A, Dieckow J. Carbon sequestration in two Brazilian Cerrado soils under no-till. Soil & Tillage Research 2006; 86(2) 237 — 245.

[50] Calegari A, Hargrove WL, Rheinheimer DS, Ralisch R, Tessier D, Tourdonnet S, Guimarães MF. Impact of long-term no-tillage and cropping system management on soil organic carbon in an Oxisol: a model for sustainability. Agronomy Journal; 100(4) 1013 — 1019.

[51] Calonego JC, Rosolem CA. Soil water retention and S index after crop rotation and chiseling. Revista Brasileira de Ciência do Solo 2011;35(6) 1927 – 1937.

[52] Maia CE. Índice S para avaliação da qualidade física de solos. (In Portuguese, with English abstract). Revista Brasileira de Ciência do Solo 2011;35(6) 1959 – 1965.

[53] Andrade RS, Stone LF. Índice S como indicador da qualidade física de solos do cerrado brasileiro. (In Portuguese, with English abstract). Revista Brasileira de Engenharia Agrícola e Ambiental 2009(4) 382 — 388.

[54] Han H, Giménez D, Lilly A. Textural Averages of Saturated Soil Hydraulic Conductivity Predicted from Water Retention Data. Geoderma 2008; 146 121 — 128.

[55] Silva RB, Dias Junior MS, Silva FAM, Fole SM. O trafego de máquinas agrícolas e as propriedades físicas, hídricas e mecânicas de um Latossolo dos cerrados. (In Portuguese, with English abstract). Revista Brasileira de Ciência do Solo 2003;27(6) 973 — 983.

[56] Castro Filho C, Henklain JC, Vieira MJ, Casão Jr R. Tillage methods and soil water conservation in southern Brazil. Soil & Tillage Research 1991;20 271—283.

Permissions

The contributors of this book come from diverse backgrounds, making this book a truly international effort. This book will bring forth new frontiers with its revolutionizing research information and detailed analysis of the nascent developments around the world.

We would like to thank Sonia Soloneski and Marcelo L. Larramendy, for lending their expertise to make the book truly unique. They have played a crucial role in the development of this book. Without their invaluable contribution this book wouldn't have been possible. They have made vital efforts to compile up to date information on the varied aspects of this subject to make this book a valuable addition to the collection of many professionals and students.

This book was conceptualized with the vision of imparting up-to-date information and advanced data in this field. To ensure the same, a matchless editorial board was set up. Every individual on the board went through rigorous rounds of assessment to prove their worth. After which they invested a large part of their time researching and compiling the most relevant data for our readers. Conferences and sessions were held from time to time between the editorial board and the contributing authors to present the data in the most comprehensible form. The editorial team has worked tirelessly to provide valuable and valid information to help people across the globe.

Every chapter published in this book has been scrutinized by our experts. Their significance has been extensively debated. The topics covered herein carry significant findings which will fuel the growth of the discipline. They may even be implemented as practical applications or may be referred to as a beginning point for another development. Chapters in this book were first published by InTech; hereby published with permission under the Creative Commons Attribution License or equivalent.

The editorial board has been involved in producing this book since its inception. They have spent rigorous hours researching and exploring the diverse topics which have resulted in the successful publishing of this book. They have passed on their knowledge of decades through this book. To expedite this challenging task, the publisher supported the team at every step. A small team of assistant editors was also appointed to further simplify the editing procedure and attain best results for the readers.

Our editorial team has been hand-picked from every corner of the world. Their multi-ethnicity adds dynamic inputs to the discussions which result in innovative outcomes. These outcomes are then further discussed with the researchers and contributors who give their valuable feedback and opinion regarding the same. The feedback is then collaborated with the researches and they are edited in a comprehensive manner to aid the understanding of the subject.

Apart from the editorial board, the designing team has also invested a significant amount of their time in understanding the subject and creating the most relevant covers. They scrutinized every image to scout for the most suitable representation of the subject and create an appropriate cover for the book.

The publishing team has been involved in this book since its early stages. They were actively engaged in every process, be it collecting the data, connecting with the contributors or procuring relevant information. The team has been an ardent support to the editorial, designing and production team. Their endless efforts to recruit the best for this project, has resulted in the accomplishment of this book. They are a veteran in the field of academics and their pool of knowledge is as vast as their experience in printing. Their expertise and guidance has proved useful at every step. Their uncompromising quality standards have made this book an exceptional effort. Their encouragement from time to time has been an inspiration for everyone.

The publisher and the editorial board hope that this book will prove to be a valuable piece of knowledge for researchers, students, practitioners and scholars across the globe.

List of Contributors

Dakshina R. Seal
Tropical Research and Education Center, University of Florida, Institute of Food and Agricultural Sciences, Homestead, FL, USA

Garima Kakkar
Fort Lauderdale Research and Education Center, University of Florida, Institute of Food and Agricultural Sciences, Davie, FL, USA

Cindy L. McKenzie
United States Department of Agriculture, Agricultural Research Services, Fort Pierce, FL, USA

Lance S. Osborne
Mid-Florida Research and Education Center, University of Florida, Institute of Food and Agricultural Sciences, Apopka, FL, USA

Vivek Kumar
Tropical Research and Education Center, University of Florida, Institute of Food and Agricultural Sciences, Homestead, FL, USA
Mid-Florida Research and Education Center, University of Florida, Institute of Food and Agricultural Sciences, Apopka, FL, USA

Joyce E. Parker, George C. Hamilton and Cesar Rodriguez-Saona
Department of Entomology, Rutgers University, New Brunswick, NJ, USA

William E. Snyder
Department of Entomology, Washington State University, Pullman, WA, USA

S.A. De Bortoli, R.A. Polanczyk, A.M. Vacari, C.P. De Bortoli and R.T. Duarte
Department of Plant Protection, FCAV-UNESP, Jaboticabal, Sao Paulo, Brazil

Hernández F.D. Castillo and Gallegos G. Morales
Universidad Autónoma Agraria Antonio Narro, México

Castillo F. Reyes
Instituto de Investigaciones Forestales Agrícolas y Pecuarias, México

Rodríguez R. Herrera and C. Aguilar
Universidad Autónoma de Coahuila, México

G.R. Mohammadi
Department of Crop Production and Breeding, Faculty of Agriculture and Natural Resources, Razi University, Kermanshah, Iran

C.O. Ehi-Eromosele and O.O. Ajani
Chemistry Department, School of Natural and Applied Sciences, College of Science and Technology, Covenant University, Canaan Land, Ota, Ogun State, Nigeria

O.C. Nwinyi
Department of Biological Sciences, School of Natural and Applied Sciences, College of Science and Technology, Covenant University,, Canaan Land, Ota, Ogun State, Nigeria

Timothy Coolong
Department of Horticulture, University of Kentucky, USA

Cezar Francisco Araujo-Junior and Júlio César Dias Chaves
Soils Area at Agronomic Institute of Paraná – IAPAR, Londrina, State of Paraná, Brazil

Benedito Noedi Rodrigues
Plant Production Area at Agronomic Institute of Paraná – IAPAR, Brazil

George Mitsuo Yada Junior
Scientific Initiation Program at Agronomic Institute of Paraná ProICI - IAPAR, Federal University Technology of Paraná – UTFPR, Londrina, PR, Brazil

Printed in the USA
CPSIA information can be obtained
at www.ICGtesting.com
JSHW011404221024
72173JS00003B/422

9 781632 396235